THE

MIND IN
THE UNIVERSE

ENCHANTED LOOM

Robert Jastrow

A TOUCHSTONE BOOK
Published by Simon and Schuster
NEW YORK

Copyright © 1981 by Reader's Library, Inc.

First Touchstone edition, 1983

Published by Simon and Schuster
A Division of Gulf & Western Corporation
Simon & Schuster Building
Rockefeller Center
1230 Avenue of the Americas
New York, New York 10020

TOUCHSTONE and colophon are trademarks of Simon & Schuster

Manufactured in the United States of America
1 3 5 7 9 10 8 6 4 2
1 3 5 7 9 10 8 6 4 2 Pbk.

Portions of this work previously appeared in the January/February and March
1981 issues of *Science Digest*

Library of Congress Cataloging in Publication Data

Jastrow, Robert, date.
The enchanted loom.

(A Touchstone book)
"Portions of this work previously appeared in the
January/February and March 1981 issues of Science
digest"—T.p. verso.
Bibliography: p.
Includes index.
1. Intellect—Evolution. 2. Brain—Evolution.
3. Human evolution. 4. Artificial intelligence.
5. Cosmology. I. Title.
BF431.J33   1983   153   82-19620
ISBN 0-671-43308-3
ISBN 0-671-47068-X Pbk.

To my
DARTMOUTH AND COLUMBIA STUDENTS
WHOSE INTELLECTUAL HONESTY AND CURIOSITY
ARE A NEVER ENDING SOURCE
OF PLEASURE

THE
ENCHANTED
LOOM

Contents

Preface

> "It is as if the Milky Way entered upon some cosmic dance. Swiftly the brain becomes an enchanted loom where millions of flashing shuttles weave a dissolving pattern, always a meaningful pattern though never an abiding one; a shifting harmony of subpatterns."
> — SIR CHARLES SHERRINGTON

This book is the third volume in a trilogy following RED GIANTS AND WHITE DWARFS and UNTIL THE SUN DIES. RED GIANTS AND WHITE DWARFS was concerned largely with the astronomical setting for human existence, and relegated the appearance of man to its concluding pages. UNTIL THE SUN DIES condensed the astronomical setting and expanded the history of life, dealing at length with the forces that shaped man into his present form. The nature of the brain was discussed, but not in detail. The new book takes up where UNTIL THE SUN DIES left off. It recaps the astronomical setting and the early history of life in the first chapter; then it focuses on intelligence and the brain: how the brain evolved, the way it works, how it balances instinct and reason, what it is evolving into.

I first became interested in these matters some years ago,

when I did two television series for CBS on the scientific background to the space program. My research for these series led me to the conclusion that while space travel has important practical applications, it also has a larger meaning in the history of life on this planet, comparable in significance to the move of the fishes out of the water 350 million years ago.

These developments broadened my research interests from astronomy to biological evolution and the history of life. One of the striking features in that history is a trend toward greater brain size and intelligence in the highest forms of life, which has persisted for many millions of years. Throughout this long interval a succession of brainy creatures has appeared, each more intelligent than its predecessors, and each serving in turn as the rootstock for a newer and still more intelligent form. We are the latest in the succession, but probably not the end of the line. Who will follow us? What will man's successor look like? My answer to this question is not the usual one. I was led to it by my work in NASA.

When I joined the space program as a young physicist, my research involved calculations in fields of science that are heavily dependent on mathematics and solving equations. This work brought me into contact with the entity called the computer. At that time, computers were little more than fast calculating machines that could add a column of figures in a fraction of a second, or integrate differential equations more rapidly than any mathematician.

In the course of some years, our research led us away from stars and planets and into more practical applications of space science. We were particularly interested in the use of satellite measurements as a tool for improving weather forecasts. Weather forecasting was a quite different problem than calculating the structure of a galaxy, or the birth and death of a star. It required a staggering amount of computations — nearly a trillion additions and subtractions for one three-day forecast. In the language of the professional, weather forecasts were "number-crunchers".

In 1975, the Goddard Institute installed the first fourth-generation computer to be put into use anywhere in the United States. It was intended to help with our weather research and to give us a start on the even tougher problem of long-range climate forecasts. By this time, the Institute's original computer - a transistorized machine of the 'second generation', as the computer experts called it — had been replaced by a machine of the third generation — the famous IBM 360. This machine weighed many tons and took up nearly the entire floor of our building. The fourth-generation computer was just as powerful as the IBM 360, and occupied only a hundredth as much space.

I stood there, looking at the giant third-generation machine purring through its sums, and then at the little fourth-generation computer off in the corner. Suddenly I became aware that powerful forces were at work. Because compactness leads to greater speed in computers*, and speed leads to better weather forecasts, as well as improvements in everything else people use computers for, computer designers are always under pressure to turn out more compact models. If these trends in computer evolution continued, a machine that occupied a floor of our Institute would soon fit into a thimble, and its circuits would be as densely packed as the electrical circuits in the human brain. If it became possible to wire those circuits so that they worked in the same way as the circuits in the brain, man would be able to create a thinking organism of quasi-human power — a new form of intelligent life.

---

*It takes pulses of electricity a certain amount of time to travel from one place to another in a computer and the smaller the distance to be travelled, the faster the machine can operate.

These thoughts captured my imagination: A non-biological intelligence, springing from the human stock, and destined to surpass its creator. I looked into the matter and studied what is known about the way in which brains work, and how computers work today and may work tomorrow. This book is the result.

Acknowledgments

Many colleagues and friends have contributed invaluable information and assistance during the development of these ideas on the brain and the computer. Dr. Paul Schneck, formerly my colleague in the Goddard Institute for Space Studies, and one of the rare and gifted individuals described in Chapter 11 who can talk to the computer in its own language, has patiently explained the principles of computer design and electronics to me in many conversations during the past several years. The discussion in the later chapters of the book owes more to him than to any other person. I am also indebted to Drs. Kenneth Korey of Dartmouth College and Robert Eckhardt of Simon and Schuster for valuable criticism of the chapters on the reptiles and mammals and human evolution, and to Dr. Michael Rampino of Columbia University for very interesting comments on the geological and paleontological evidence relating to the extinction of the dinosaurs. Dr. Eckhardt gave the manuscript a careful editorial review and made numerous suggestions that improved the clarity of the exposition. Dr. Steven Ungar of the Goddard Institute provided helpful information on the processing of visual images by computers. Dr. T. N. Wiesel of the

Harvard Medical School was kind enough to read the chapter on the brain and vision. His suggestions substantially improved the accuracy of this section.

Doris Cook has been a valued partner in many stimulating conversations on both brains and computers. Her contributions are numerous and important, and appear in every chapter. The book owes a great deal to the clarity of her thought. My mother, Marie Jastrow, read the manuscript carefully and made many suggestions that improved its readability for the general reader. Her criticisms were particularly helpful in the preparation of the charts and illustrations, where my scientific training proved a handicap. Susan Tufts contributed in a major way to the editing of the manuscript, and also provided essential support in all aspects of the production of the manuscript for the printer. Susan Messer joined us later in the project and played an important role in the final stages of editing as well as the preparation of charts and illustrations. Four talented young artists — Jane Svoboda, Andrea Calarco, J. B. McKoy III and Marrin Robinson — did most of the drawings. Jack Hall prepared the charts in chapters 2 and 10.

I should like to reserve a final and very special word of appreciation for Erwin Glikes, my editor at Simon and Schuster, whose contributions went beyond the limits of the normal author-editor relationship. They strengthened the book enormously, particularly in the chapters on the computer, which were the most difficult of all to write.

THE
ENCHANTED
LOOM

1·Across the Threshold of Life

Scientific discoveries of the last decades have created a new explanation for the appearance of man on the earth. In the scientist's version of Genesis, as in the Bible, the world begins with the dazzling splendor of the moment of creation. Few astronomers could have anticipated that this event — the sudden birth of the Universe — would become a proven scientific fact, but observations of the heavens through telescopes have forced them to that conclusion.

The first scientific indication of an abrupt beginning for the world appeared about fifty years ago. At that time, American astronomers, studying the great clusters of stars called galaxies, stumbled on evidence that the entire Universe is blowing up before our eyes. According to their observations, all the galaxies in the Universe are moving away from us and from one another at very high speeds, and the most distant are receding at extraordinary speeds of hundreds of millions of miles an hour. This discovery led directly to the picture of a sudden beginning for the Universe; for if we retrace the movements of the expanding galaxies backward in time, we find that at an earlier time they must have been closer together than they are today; at a still earlier time, they must have been still closer together; and if we go back far enough in time, we find that at a certain critical moment in the past, all the galaxies in the Universe were packed together into one dense mass, at an enormous pressure and temperature. Reacting to this pressure, the dense, hot matter must have exploded with incredible violence. The instant of the explosion marked the birth of the Universe.

The seed of everything that has happened in the Universe was planted in that first instant; every star, every planet and every living creature in the Universe came into being as a result of events that were set in motion in the moment of the cosmic explosion. It was literally the moment of Creation.

When did it happen? When did the Universe explode into being? Calculations based on the present positions of galaxies show that this great event occurred twenty billion years ago. Twenty billion years is the age of the Universe, according to the astronomers. This is a very long time. The sun and the earth have only existed for four and a half billion years, and life has been on the earth for even less time than that. Humanity has existed on our planet for only one million years, which is less than one ten-thousandth as long as the age of the Universe. These intervals of time begin to place man's existence in a cosmic perspective.

Astronomers have been working on bits and pieces of this remarkable story for years, little dreaming that they were making contributions to a new and updated version of the Book of Genesis. Many details of the scientific account differ from those in the Bible; in particular, the age of the Universe appears to be far greater than the 6000 years of the Biblical Genesis; but the essential feature is the same in both stories: There was a Beginning, and all things in the Universe can be traced back to it.

The astronomical proof of a Beginning places scientists in an awkward position, for they believe that every effect has a natural cause, and every event in the Universe can be explained by natural forces, working in accordance with physical law. Yet science can find no force in nature that might account for the beginning of the Universe; and it can find no evidence that the Universe even existed before that first moment. The British astronomer E. A. Milne wrote, "We can make no propositions about the state of affairs [in the beginning]; in the Divine act of creation God is unobserved and unwitnessed."

But when it comes to explaining the events that took place after the Beginning, scientists feel more confident. With the aid of telescopes and other instruments, they have reconstructed the long chain of events, extending over billions of years, in which the gases of the newly created Universe were slowly transformed into stars, planets and conscious life.

The narrative pieced together by the astronomers starts immediately after the Creation, when the Universe was very hot and filled with radiant energy. Driven by its internal pressures, the hot, young Universe expanded rapidly, and cooled as it expanded. By the time the Universe was about one million years old, its temperature had fallen to a few thousand degrees, and electrons collected around nuclei to form the first atoms. From that point on, the primordial matter of the Universe consisted of gaseous clouds of hydrogen and helium atoms, drifting through a vast and dark space.

With the passage of time, the matter of the Universe cooled further. When the Universe was about one billion years old, clouds of its gases were cool enough so that knots of condensed matter began to appear in the swirling gases. Clouds of atoms came together to form galaxies; within the galaxies, smaller clouds condensed into stars. Around some stars, planets formed.

For many billions of years, stars and planets continued to form in the surging gases of the Universe. Stars are still forming today in the heavens — spheres of hot gas, radiating heat and light, and soon to be suns like ours. The sun formed in this way four and a half billion years ago, freshly condensed out of the swirling mists of the Galaxy.

Around the newborn sun, smaller knots of gas and dust condensed to form the earth and its sister planets. Exactly how that happened is one of the mysteries of science. We understand the life cycle of the stars quite well, because we see them in the sky today in many different stages of their lives — young, middle-aged, and old. We can even see stars being born in the sky at this very moment. But no one has ever seen a planet being born in the sky. The only planets we can see, even through the largest telescopes, are the earth's sister planets in our own solar system,* and none of these is newborn. Each planet in the solar system is the same age as the earth; each was formed four and a half billion years ago, along with the earth, when the solar system came into being.

Yet somehow it happened; somehow, mysterious forces, acting on the atoms of gas and grains of rock and iron that circled the young sun, swept these materials together to form the nine planets, their forty-odd moons, and uncounted comets and asteroids. As the earth grew, its gravity increased in strength, and drew bits of matter to it from the surrounding space with greater and greater force. Colliding with the earth's surface, these fragments of rock liberated vast amounts of heat, which melted the outer layers of the planet. In the end, the earth circled in its orbit around the sun,

*Planets cannot be seen through telescopes unless they are inside our own solar system. We hope to be able to see planets in other solar systems for the first time in 1985, when a large telescope will be carried into orbit by the Space Shuttle.

newborn, hot as a furnace, and devoid of living organisms.

Now another mystery interrupts the scientist's story. According to the fossil record, simple kinds of life appeared on the earth at some point during the first billion years of its existence. Where did these living organisms come from? Since the earth's surface was too hot to bear life at the start, this life must have appeared on our planet later. Either it was placed here by the Creator, or it evolved out of nonliving molecules in accordance with the laws of chemistry and physics. There is no third way; it must have been one or the other.*

Scientists have no proof that life was not the result of an act of creation, but they are driven by the nature of their profession to seek explanations for the origin of life that lie within the boundaries of natural law. They ask themselves, "How did life arise out of inanimate matter? And what is the probability of that happening?" And to their chagrin they have no clear-cut answer, because chemists have never succeeded in reproducing nature's experiments on the creation of life out of nonliving matter. Scientists do not know how that happened, and, furthermore, they do not know the chance of its happening. Perhaps the chance is very small, and the appearance of life on a planet is an event of miraculously low probability. Perhaps life on the earth is unique in this Universe. No scientific evidence precludes that possibility.

But while scientists must accept the possibility that life may be an improbable event, they have some tentative reasons for thinking that its appearance on earthlike planets

---

*A theory popular in the nineteenth century, and revived recently, proposes that life came to the earth from some distant star, carried here in the interiors of spores resistant to the cold and lethal radiation of space. However, this theory does not offer a truly different explanation of the origin of life; it only postpones the question to another time and place.

MAN'S DESCENT FROM THE STARS.
Stars and planets began to form nearly 20
billion years ago. Countless stars were formed
before the sun and earth were born; many
can still be seen forming in the sky today.

This photograph, taken at the 120-inch
telescope at Lick Observatory, shows a region
in our Galaxy called the Serpens nebula,
which is rich in newly forming stars. The
small, dark regions scattered across the
photograph are clouds of dense matter in the
process of condensing into stars. The bright
regions in this beautiful nebula are thinner
clouds of hydrogen made luminous by ultra-
violet rays absorbed from the many hot,
young stars embedded in the nebula.

Man can trace his origins back to a cloud
like this, that condensed out of the gas and
dust of the Galaxy 4.6 billion years ago. According to the scientist's reconstruction of
the story of Genesis, every grain of rock,
every molecule of water, and every living
thing on the surface of the earth is descended
from the atoms of that parent cloud.

is, in fact, fairly commonplace. These reasons do not constitute proof, but they are suggestive. Laboratory experiments show that certain molecules, which are the building blocks of living matter, are formed in great abundance under conditions resembling those on the earth four billion years ago, when it was a young planet. Furthermore, those molecular building blocks of life appear in living organisms today in just about the same relative amounts with which they appear in the laboratory experiments. It is as if nature, in fashioning the first forms of life, used the ingredients at hand and in just the proportions in which they were present.

Moreover, nature's experiments on the origin of life seem to have come to fruition in a rather short time. This fact suggests that the experiments may have been easy, and the chances of success fairly high. According to the fossil record, relatively complicated organisms like bacteria already existed when the earth was only one billion years old. Although a bacterium seems like a simple kind of life to us, it is a quite complex chemical factory, whose existence depends on the simultaneous manufacture of several thousand different kinds of chemicals. Bacteria are far more advanced than those simple creatures that first wriggled across the threshold of life on the earth.

If bacteria already existed when the earth was one billion years old, a long period of evolution must have preceded their appearance, in which the chemical machinery that makes up the business of life for a bacterium was slowly being worked out and improved. This implies that the threshold of life itself must have been crossed far earlier — perhaps when the earth was only a few hundred million years old, or even younger. A few hundred million years is not a long time for such an important experiment; if the experiment succeeded as quickly as that, the probability of its success must have been fairly high.

In any case, life did appear on the earth when it was still a very young planet. Nature set to work immediately to improve those first simple organisms, but no trace of that early stage in the development of life has been preserved.

The record of the rocks contains very little, other than bacteria and one-celled plants until, about a billion years ago, after some three billion years of invisible progress, a major breakthrough occurred. The first many-celled creatures appeared on earth. Man is a many-celled animal, and one of the products of that evolutionary event.

The fossil record contains the remains of these many-celled organisms. These were primitive, soft–bodied animals; nonetheless, they were a great advance over single cells like the bacterium. A many-celled animal is a colony that contains enormous numbers of single cells bonded together. It is physically strong, and far less vulnerable to hostile forces in the environment than a single cell would be; furthermore, individual cells in the colony quickly acquire specialized functions — cells in the gut for digesting food, tough cells on the surface to make the skin, and sensitive cells at the forward end for the beginnings of the eye, the nose and the ear. These improvements greatly enhance the prospects of survival for the individual; they are the plan on which all the higher forms of life are constructed.

During the next half billion years or so, very little happened; at least, little that is preserved in the fossil record. Then 600 million years ago, another great advance occurred. The fossil record shows that at that time the first hard-bodied creatures — animals with external skeletons — appeared on the earth. These were the ancestors of the clam, the starfish, the lobster, and the insect. They were a simple kind of life; they still lacked a brain; but at least they possessed body armor. Now the pace of evolution quickened; in the relatively short interval of 100 million years that followed, some hard-bodied animals evolved into a still newer kind of creature called a vertebrate, that possessed an internal skeleton with a backbone — a flexible, hinged chain of girders against which the muscles of the body could pull for greater strength and agility.

In a relatively few tens of millions of years the vertebrates evolved into better animals — the first fishes. The fishes possessed a first-class skeleton with nicely articulated bones in fin as well as spine. They also possessed a brain. It was a very small brain, but it was the first one that had existed on the earth up to that time.

The fishes appeared in the waters of the earth about 450 million years ago. Prior to that time, all life had been confined to the water, and the land was as barren and bleak as the surface of the moon. Now life's conquest of the land began. The plants came first. Slowly, fingers of green worked their way inland from the shore probing for moist places. Some tens of millions of years later the insects followed, lured by the abundance of vegetation on the land. The insects were the first animals to invade the land, and their life was relatively free from predators at first. The freedom of the insects did not last long, because 50 million years later, in a

THE BRAIN OF THE FISH. The brain of the fish, shown in an enlarged view at the far right, represents an early stage in the evolution of the brain in backboned animals. The fish's brain is divided into three compartments: a front section for smell, a middle section for vision, and a rear section for balance.

The receptors for smell lead directly from the fish's nose into the front compartment, or smell brain. The optic nerve carries information from the fish's eyes into the vision brain (shaded area), behind the smell brain. The vision brain is the largest region in the fish's brain, and is its most important single source of information. When the vision brain receives a report from the fish's eyes, it reacts immediately according to a simple, unvarying program of behavior. This is an unthinking brain; because of its small size, it has little room for flexible responses or coordinating information from several senses.

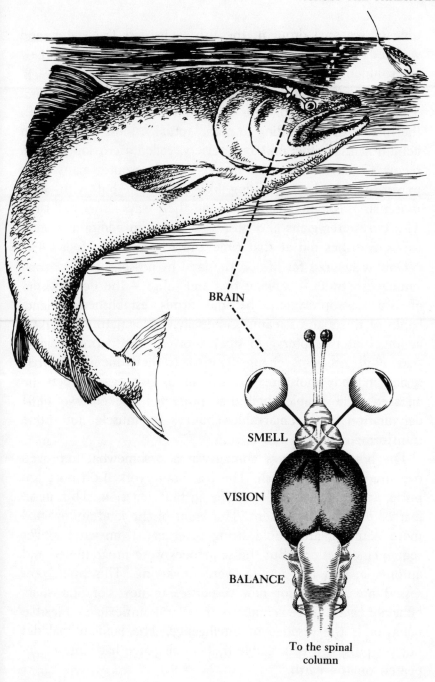

BRAIN

SMELL

VISION

BALANCE

To the spinal
column

time of seasonal drought, the fishes also left the water, and waddled out on stumpy fins to taste the pleasures of the new environment.

The migration of the fishes onto the land occurred about 350 million years ago. This migration is another highlight in the history of life leading to man. Actually, most fishes stayed behind at that time; only one variety, called a crossopterygian, seems to have made the move. The crossopterygians were powerful animals, something like a salmon, but with thick muscular fins, suitable for pushing their way across the land. The crossopterygians also possessed lungs, in addition to gills, as many fishes did at that time and some still do today.

Doubly favored for life on the land by possession of two advantageous traits — walking fins and lungs — the descendants of the crossopterygians became firmly established on the banks of rivers and streams. Gradually, under nature's pruning action, the crossopterygian stock evolved. The permeable fish skin of its ancestors changed into a leathery membrane that sealed in body moisture; the gelatinous fish egg came to be encased in a pliable shell that protected the embryo from dehydration; and remarkable changes in muscle and bone transformed the fin into a limb.

The brain of the new animal was also somewhat improved over the brain of the fish. The fish brain worked on pure impulse; stimulus was followed by instant reaction. This mind existed only in the present. The brain of the land animal had more neurons, and could string together a somewhat longer sequence of actions. But the sequences were programmed and automatic, and the actions were unvarying. This brain possessed no capacity for new responses to new conditions; its behavior was more complex, but still inflexible. Flexible behavior is the essence of intelligence. The land animal did not yet possess this desirable trait; intelligence had not yet appeared on the earth.

As always in evolution, these changes in body and brain proceeded at a relatively slow pace, with imperceptible changes from one generation to the next. After some 25 million generations, a transitional form appeared, that lived part of its life on the land and part in the water. This was the amphibian, ancestor of the frog and the toad, still fishlike in eggs and skin. Finally, after 50 million generations, the improvements were complete. At the end of that long interval an entirely new animal existed, completely emancipated from the water, in which the outlines of its fishlike ancestor were only faintly visible. The new animal was the reptile.

The reptiles appeared on the earth 300 million years ago. Like the insects before them, they held undisputed mastery over the land for a time, with no enemy except their own kind. They flourished and gave rise to many great lines of evolution — lizards and snakes, turtles, crocodiles, dinosaurs, birds — and the mammals.

All these ancient forms of life — the fishes, the early reptiles, the dinosaurs, and the birds — were relatively unintelligent, and their descendants remain so today. But the mammals were different; the story of the evolution of intelligence is really *their* story. The appearance of the mammals marked the first great step in the evolution of the brain.

FROM FISH TO MAN. The crossopterygians (left) were fish with lungs and stumpy fins that left the water to explore the land 350 million years ago, in a time of seasonal drought. These fish are the ancestors of every four-limbed, backboned animal on the face of the earth.

The lower illustration shows how the bones in the fin of the fish evolved into the bones of the human arm and hand. The fossil record documents many intermediate stages in this transition. Much of the human skeleton can be traced back, bone by bone, to the skeletons of our fish ancestors.

SHOULDER

UPPER ARM (HUMERUS)

ELBOW

FOREARM

WRIST

TODAY

200 MILLION YEARS AGO _____ THE FIRST MAMMALS

300 MILLION YEARS AGO _____ THE FIRST REPTILES

450 MILLION YEARS AGO _____ THE FIRST FISHES

600 MILLION YEARS AGO _____ THE FIRST HARD-BODIED
ANIMALS

1 BILLION YEARS AGO _____ MANY-CELLED ANIMALS

3.5 TO 4 BILLION YEARS AGO _____ LIFE APPEARS
BACTERIA AND SIMPLE
PLANTS

4.6 BILLION YEARS AGO _____ THE SUN AND EARTH APPEAR

15 BILLION YEAR GAP _____ STARS AND PLANETS FORM
THROUGHOUT

1 BILLION YEARS LATER _____ STARS BEGIN TO FORM

20 BILLION YEARS AGO _____ THE UNIVERSE EXPLODES
INTO BEING

30

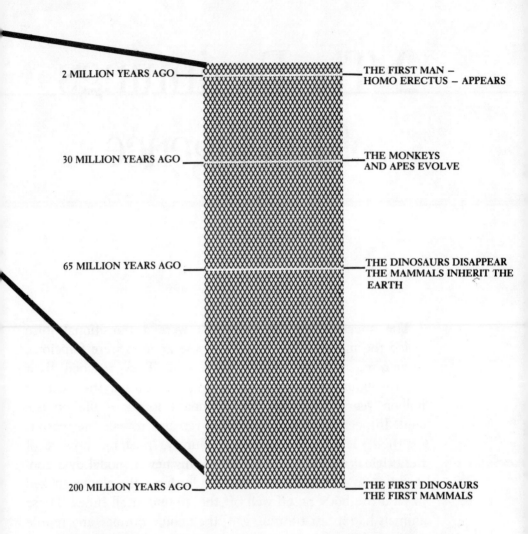

2 MILLION YEARS AGO	THE FIRST MAN – HOMO ERECTUS – APPEARS
30 MILLION YEARS AGO	THE MONKEYS AND APES EVOLVE
65 MILLION YEARS AGO	THE DINOSAURS DISAPPEAR THE MAMMALS INHERIT THE EARTH
200 MILLION YEARS AGO	THE FIRST DINOSAURS THE FIRST MAMMALS

THE COLUMN OF TIME. The great events in the history of the earth and life on the earth are shown here. The bottom of the left-hand column represents the moment of creation. Stars and planets began to form after one billion years, and have continued to form to the present day. The sun and earth condensed out of gas and dust 4.6 billion years ago. Less than a billion years later, life appeared on the earth. Little happened for several billion years thereafter; then the pace of evolution quickened.

The history of intelligence is confined to the narrow slice of time — the last 200 million years — in which the mammals have existed on the earth. The right-hand column shows the evolutionary highlights in this final period. Man appeared after 99.9 per cent of the history elapsed.

2·The Beginnings of Intelligence

The ancestors of the mammals were a transitional form called the mammal-like reptiles. These animals were ferocious, numerous, and enormously successful. They reached their zenith about 250 million years ago, and during the next 50 million years they were the dominant forms of life on the land. Inspection of their fossilized remains reveals the reasons for their success. The sprawling, clumsy, "push-up" posture of the original reptile was replaced in this newer model by a construction in which the legs were underslung, the elbows tucked in, and the body raised well off the ground at all times. These animals had a fast running gait; they could outrace any reptile in their day.

The skeletons of the mammal-like reptiles suggest changes in their behavior as well. The reptile lies on its belly most of the day, conserving energy; its coldblooded metabolism does not afford it the luxury of running around continuously. The mammal-like reptiles, trotting across the ground much of the time, would have expended energy at a greater rate. Even standing still required more energy than lying about on the ground with their reptile cousins. In a standing animal, the muscles of the leg and trunk constantly tense and relax, working at keeping the body upright, burning fuel and releasing heat. That constant activity, producing heat even at rest, must have increased the body temperature of the mammal-like reptile. This was the first step toward the warmblooded metabolism that characterizes the modern mammal.

Warmbloodedness carries a bonus. When dusk falls and the temperature of the air drops, the coldblooded reptile, losing body heat, becomes drowsy and retires for the night. The mammal-like reptile, with traits of warmbloodedness, can stay up later and wander about, snapping up its torpid reptilian cousins.

It is possible that the mammal-like reptiles also possessed rudimentary traits of the parental love and care that distinguish mammals today. Some evidence for this view comes from the discovery of an infant mammal-like reptile lying close to the skeleton of an adult of the same kind. Parental care would have been a radical change from the behavior of the older kinds of reptiles. Most reptiles display no parental feelings; they lay their eggs and leave them, and once the hatchlings are on their own, the parents will eat their young if they can catch them.* Animals that eat their young would seem to be doomed to early extinction, but reptiles lay large numbers of eggs to compensate for their ungracious family manners.

The duckbilled platypus of Australia provides additional evidence that parental care developed early in the line of the mammal-like reptiles. This bizarre creature, half mammal and half reptile, is a living fossil, and represents the most primitive and reptile-like stage in the development of the mammals. Some scientists believe that the platypus split off from the

---

*The crocodile is an exception. Most snakes, lizards and turtles display no sense of obligation toward their offspring.

main line of mammal evolution 200 million years ago, and is the only living direct descendant of the mammal-like reptiles. Although the platypus lays eggs like a reptile, walks like a reptile, and has a reptile's plumbing, it has the mammal's coat of fur, a mammal's temperature controls for warmbloodedness — and it nurses its young.

The traits of parental care that developed in the mammalian line of evolution were accompanied by a decrease in the size of the litter. This was a new strategy for maintaining the species, and quite different from any that had been practiced up to that time. In place of dozens of eggs laid by the reptile, only a few young appear in a mammal's litter, but a great deal of care and attention is lavished on each infant. In that innovation lie the seeds of the pair bond, the family tie, and much that we consider human.

Parental care and warmbloodedness go together. Among warmblooded animals, in whom a constant body temperature is essential for survival, the newborn young are particularly vulnerable to death by freezing because of their small size and nakedness. The mammal-like reptile parent that stayed with its young, and kept them warm, was more likely to see the litter survive; and, by the laws of inheritance, its young were likely to carry the desirable traits of parental care into the next generation, and bestow them on their offspring. Indifferent parents among the mammal-like reptiles left fewer offspring, and their traits tended to disappear from the population. Good parenthood and a warmblooded metabolism probably entered the history of life together, with the advent of the mammal-like reptiles.

These diverse traits of body and behavior worked together; each intensified the others. All were experimental; none had been tried before by nature; all turned out to be advantageous. The mammal-like reptiles flourished, and in the course of some millions of years they became the rulers of the

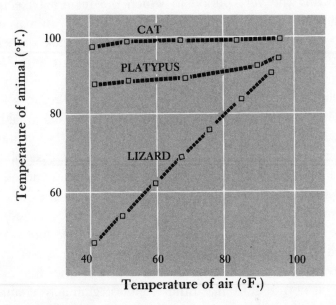

WARMBLOODEDNESS. The meaning of warmbloodedness is illustrated by this graph, which shows how the body temperatures of different kinds of animals change with the temperature of their surroundings.

The cat is a typical warmblooded mammal, with many special ways of keeping its body temperature constant. For example, sweat glands or panting cool a warmblooded animal when the air is hot, and the blood vessels under the skin dilate to carry body heat to the surface, where it is lost to space. Because of these defenses, the temperature of the cat scarcely changes when the temperature increases from 40°F to 110°F.

The lizard — a typical reptile — is coldblooded and lacks the cat's defenses against temperature change. Its body temperature is always closely equal to the temperature of the outside air.

The duck-billed platypus — a link between the mammals and the reptiles — controls its body temperature more effectively than the lizard but not as well as the cat.

earth. They included many varieties of plant-eaters, meat-eaters who ate the plant-eaters, some animals the size of a squirrel, and others the size of a moose. They were the most successful animals of their time.

But during the very period in which the mammal-like reptiles enjoyed their supremacy, another kind of animal was evolving, descended from the same ancestral stock of belly-dragging reptiles, but modelled along very different lines. The new kind of animal was the dinosaur.

The line of evolution leading to the dinosaur began about 225 million years ago, during the heyday of the mammal-like reptiles. In this line of evolution speed and agility, so valuable for survival, were again of the essence. As in the case of the mammal-like reptiles, these desirable traits were achieved in the dinosaurs by a change in posture, in which the legs no longer sprawled outward to either side but were tucked in, with the body raised off the ground. Still another innovation appeared in the dinosaur line, that led to an even better body design than that of the mammal-like reptiles. The dinosaurs, four-footed at the start like other reptiles, gradually changed to a two-legged posture. As people and birds do today, they walked and ran on their hind legs. Their hind limbs became strong and muscular, and gave them additional speed, while

THE ANCESTRAL DINOSAUR. This small reptile, the ancestor of the dinosaurs, lived about 225 million years ago.

their forelimbs were freed for grasping their prey with cruel talons.

These were the early dinosaurs. They were not very large; in appearance they resembled a scrawny, featherless rooster, with the long tail of a lizard attached. A jaw with a wide gape, filled with rows of razor teeth, completed the picture of a fierce and effective little carnivore.

Were the dinosaurs warmblooded? That might seem to be necessarily so, at least for the smaller, more agile dinosaurs. The lithe construction of those reptiles suggests an active lifestyle, and a rapid expenditure of energy. Usually these traits require a warmblooded metabolism. On the other hand, the fossilized teeth of Tyrannosaurus rex, a large carnivorous dinosaur, show annual growth rings similar to tree rings that reflect seasonal changes in the average body temperature of the animal. These rings are normally found only in coldblooded animals, whose average body temperature varies during the year. The growth rings in the teeth of Tyrannosaurus suggest that this dinosaur, at least, was coldblooded.

DINOSAUR TEETH. These photographs show cross sections of the teeth of the carnivorous dinosaur Tyrannosaurus (left) and a plant-eating dinosaur that lived at the same time. The rings in the teeth are believed to be annual growth marks, indicating that these were coldblooded animals.

Students of dinosaur biology are divided on the issue of the warmblooded dinosaurs, but a consensus seems to be emerging that medium-to-large-sized dinosaurs probably were neither coldblooded nor warmblooded, but had a unique metabolism of their own. The question may depend on what kind of heart the dinosaur had. The four-chambered heart of the mammal and the bird, with its double pump, is probably necessary for a high rate of energy expenditure. Modern reptiles, which are coldblooded, have a two-chambered heart. If the dinosaurs had a two-chambered heart, they probably were cold-blooded also. However, the soft parts of the dinosaur's body have not been preserved in the fossil record, and no one knows whether these reptiles had an old-fashioned heart or the more modern variety. It is possible that the controversy over dinosaur metabolism will never be settled.

Dinosaurs are thought of as very big beasts, but in the beginning all dinosaurs were small. As they prospered, they grew larger. In any given line of evolution, animals tend to increase in size from one generation to the next, other things being equal, because size provides security from attack, and thereby enhances the individual's prospects for survival. In each generation, the largest animals are the ones most likely to survive and produce offspring; their progeny, by the rules of inheritance, also tend to be larger than average; and so the tendency to increased size continues, until the limits of the available food supply are reached.

The only unusual feature in the history of the dinosaurs is the fact that just as they started to evolve, the earth entered into a period of mild climate and abundant food that lasted for an unusually long time. This circumstance permitted the growth of the dinosaurs to continue for more than 100 million years, a greater interval than any other period of uninterrupted good weather in the recent history of the earth. The giants of the land were the product of that unprecedentedly long period of good living and sustained growth.

During the course of those many millions of generations of prosperity, the descendants of the primordial dinosaurs — the lean, hungry chicken-lizards — lost their graceful forms. Some became bloated plant-eaters, and dropped back onto all fours again, forced downward by their weight. Supersaurus is the largest known of these, and the largest land animal that ever lived. In this dinosaur, a 100-ton bulk was supported on four massive, columnar legs, with spongy pads for feet. Supersaurus and its relatives were the reptilian equivalent of the cow, but each weighed as much as a herd of elephants.

Other dinosaurs remained carnivorous, and also grew larger. Eventually they evolved into fierce predators like Tyrannosaurus rex, two stories high, with thighs twelve feet round and 60 sharp teeth, set in row on row within four-foot jaws. Tyrannosaurus and his kin were the most terrifying animals that the world has ever seen.

The mammal-like reptiles were no match for the carnivorous dinosaurs. Once the dominant forms of life on the land, they were reduced, in the course of some tens of millions of years, to a small minority. They declined in size as well as numbers, perhaps because in every generation the largest and most conspicuous were picked off and eaten, and only the smallest varieties survived. From powerful animals the size of a bear, a wolf, or a dog, the descendants of the mammal-like reptiles shrank to the size of a cat, and then to small creatures the size of a rat or a mouse. By 180 million years ago, these rat-sized animals were the sole survivors of the once mighty tribe of the mammal-like reptiles.

But the furtive, ratlike creature was no longer a mammal-like reptile; it was the first true mammal. It was small, but the seeds of greatness were in it.

SPRAWLING REPTILES. Seymouria (left), closely related to the first reptiles, has been reconstructed from a fossil skeleton (below) found in Seymour, Texas. The bones in the skeleton show the transition from the amphibians to the reptiles. Seymouria represents the link between the two kinds of animals.

The sprawling, belly-dragging posture of Seymouria, characteristic of reptiles, is a legacy from fish ancestors whose fins angled straight out to either side. The limbs of these animals did not carry the body; they alternately pushed and dragged it.

These were coldblooded and sluggish creatures; they not only lacked the muscle and bone needed for a fast gait; they also lacked a warmblooded metabolism and continuous energy supply needed to fuel the active life style.

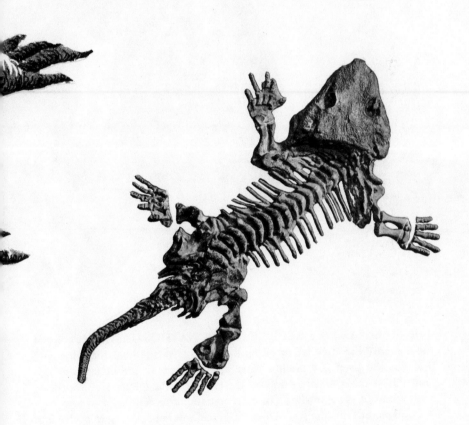

THE MAMMAL-LIKE REPTILE CYNOGNATHUS. The relative grace of the mammal-like reptiles is exemplified by Cynognathus, the "dog-jaw" (*above*), who lived 225 million years ago. Changes in the skeleton of this animal have raised its body well off the ground, although its posture still retains a hint of the sprawling gait characteristic of its reptile ancestors.

Cynognathus was the size of a wolf. Its teeth were beginning to evolve into the canines and incisors characteristic of modern mammals, and used for biting and tearing the prey into small pieces. The change in teeth was important, because it enabled the mammal-like reptile to digest its food rapidly and

gain the quick energy needed for an active lifestyle. These teeth are another indication that the mammal-like reptiles were evolving in the direction of warmblooded animals.

Cynognathus, with its mixture of lizard-like and dog-like traits, is an example of "missing links" that are no longer missing. Many such "missing links" have been found. One is a creature resembling a lizard with teeth and feathers, believed to be a half-way stage in the evolution of the birds out of the reptiles. These long-dead animals are interesting because they demonstrate the continuity of evolution.

A GALLERY OF MAMMAL-LIKE REPTILES. The mammal-like reptiles are relatively unfamiliar creatures in the history of life. However, 250 million years ago they were the most successful animals on the earth. As many varieties of mammal-like reptiles existed at that time as dinosaurs later on, or mammals today. Not all were large and ferocious like Cynognathus; some were small, like the woodchuck-sized dicynodont or "dog-tooth" (*top*), the timid-looking dromasaur (*middle*), and the bizarre-faced rhynchosaur (*bottom*).

Despite their great success, which lasted for 50 million years, the mammal-like reptiles lost out eventually to the dinosaurs. They dwindled in number and variety, and only one or two kinds survived. The survivors were the first mammals.

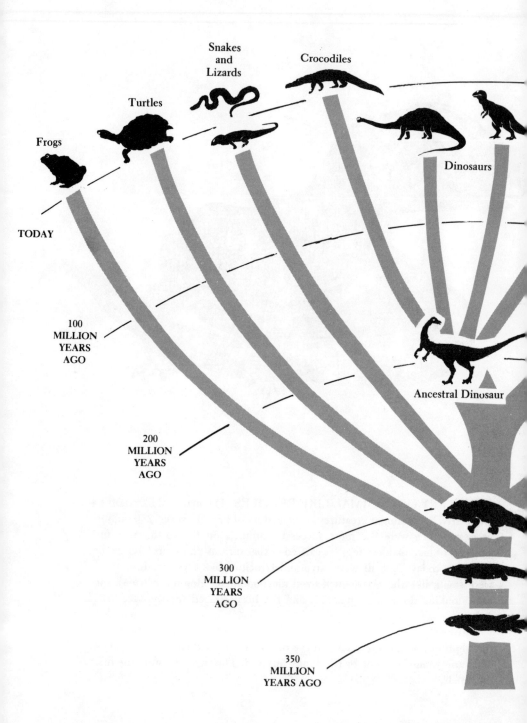

Frogs

Turtles

Snakes
and
Lizards

Crocodiles

Dinosaurs

TODAY

100
MILLION
YEARS
AGO

200
MILLION
YEARS
AGO

Ancestral Dinosaur

300
MILLION
YEARS
AGO

350
MILLION
YEARS AGO

46

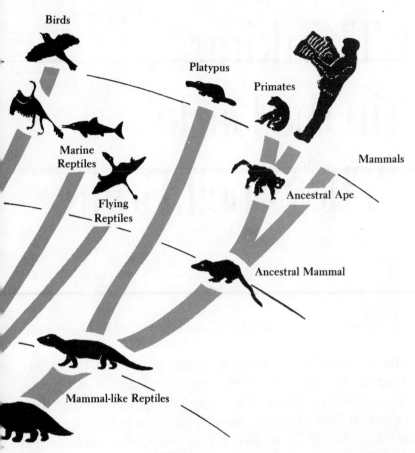

Birds

Platypus

Primates

Marine
Reptiles

Mammals

Flying
Reptiles

Ancestral Ape

Ancestral Mammal

Mammal-like Reptiles

Ancestral Reptile

Amphibian

Crossopterygian

THE EVOLUTION OF THE BACKBONED ANIMALS.
The fossil record indicates that every backboned animal on the
land and in the water can trace its ancestry to a kind of fish
called the Crossopterygian, which lived 350 million years ago.
Out of the Crossopterygian stock came the amphibians —
represented today by the frog and its cousins — and the reptiles. The reptiles arose 300 million years ago. By 200 million
years ago, the reptile line had split into several branches. One
branch gave rise to the snake, lizard, turtle, and other modern
reptiles. Other branches evolved into dinosaurs, birds, and
mammals including man. If the fossil evidence were not in
hand, it would be hard to believe that man and the dinosaur
are distant cousins — products of two lines of evolution from a
single reptilian ancestor.

3·Thinking
in the Dark:
The Smell Brain

The existence of the early mammals is revealed to us only through a few fossilized teeth and fragments of bone, but these are enough to tell their story. We can guess that they were active little animals, warmblooded, probably furry, and bore their young alive. Their eyes were very small. Their snouts were long, implying a well developed sense of smell, and the bones in the skulls indicate that they had better hearing than their mammal-like reptile forebears.

Small eyes, and a good nose and ears — they suggest an animal that lives by night; snuffling through the litter of leaves on the forest floor. The early mammals were probably nocturnal; they foraged in the dark, and kept out of sight during the day. That explains how these relatively weak and defenseless animals survived during the long reign of the dinosaurs. The mammals lived in the time of the dinosaurs, but not at the same hour.

The skulls of the first mammals reveal another important fact. These little creatures had relatively large brains. In proportion to body weight, their brains were five times larger than the brain of Tyrannosaurus, and twenty times larger than the brains of the even dimmer-witted herbivorous dinosaurs on whom Tyrannosaurus fed. The little mammal was an intelligent animal — more intelligent than any creature that had evolved on our planet until that time.

Of course, in actual size the brains of the mammals were far smaller than the brains of the dinosaurs. But the size of a brain is not as important as the *ratio* of its size to the size of the body. This ratio, averaged over all the members of a species, provides an accurate indication of the intelligence of the species.

The reason why brain size alone may be misleading is that a part of every animal's brain is used for the control of its body. This section of the brain is filled with circuits that are connected up in a kind of telephone switching exchange, receiving signals from the body and sending out messages in return. The larger the body, the larger the part of the brain that must be used for this purpose. Nearly all the available space in the dinosaur's small brain was taken up with circuits for control of its huge body; there was little room for memory, or planning, or learning by experience in that brain. But the mammal's brain, large in proportion to its body, had grey matter available for the storage of memories and for thinking, planning, and a flexible response to changing conditions. The small but relatively brainy mammal must have possessed these mental traits to a far greater degree than its brawny dinosaur contemporaries.

It is interesting to compare the brain of a very large dinosaur with the brain of an equally large modern mammal like the whale. The largest dinosaurs weighed as much as 100 tons. Whales also weigh as much as 100 tons and are, as the dinosaurs were in their time, the largest animals alive today.

AN ANCESTRAL MAMMAL. This early mammal, known as *Morganucodon* or "Morning-tooth," lived 200 million years ago. Small eyes and a long nose indicate that this animal was nocturnal and survived by smell rather than sight.

The brain of a large whale is a huge mass of grey matter, nearly a foot and a half across, that weighs about 20 pounds. The possessor of this mammoth brain is an intelligent animal. Some whales have a remarkable memory capacity; they can memorize a complex whalesong that goes on for hours, and repeat it note for note a year later.

The brains of the largest dinosaurs, on the other hand, such as Supersaurus, were only the size of an orange, and weighed about half a pound. Yet that small amount of grey matter had to exercise control over the same 100-ton bulk that is commanded by the 20-pound brain of the largest whales.

Scientists who specialize in the study of brains and intelligence have plotted charts of brain weight against body weight for many kinds of animals. They find that when the ratio of brain weight to body weight is as small as it was in Supersaurus, the behavior of the animal is stereotyped, automatic and unintelligent. The reason is clear: a large body has many large muscles, and needs many nerve fibers for its coordination. When that large body is controlled by a small brain, every neuron in this brain must be used to move the body through its basic survival routines: find food! flee from the predator! and so on.

Supersaurus was not an unusually stupid dinosaur, and dinosaurs were not unusually stupid reptiles. In fact, dinosaurs had normal intelligence for reptiles. Of course, there was a spread in braininess among the dinosaurs. But the same is true among modern mammals; plant-eaters like the cow are among the least intelligent mammals, while alert carnivores like the wolf are among the most intelligent. However, the dinosaurs as a group, were generally less intelligent than the early mammals as a group. This held then, and still holds today, all the way up and down the scale of sizes. A little lizard, for example, has a considerably smaller brain than a chipmunk of the same size, and displays a far less flexible reper-

SIZES OF HUMAN BRAINS. This table lists approximate weights of the brains of several famous men. For comparison, the average human brain weighs three pounds. The ratio of brain size to body size is a good measure of the *average* level of intelligence in a population of animals. However, as the table shows, the size of the *individual* human brain is not a good indicator of human intelligence. Victor Hugo had an enormous brain that weighed five pounds, while the brain of Anatole France, a writer of comparable stature, weighed only two pounds. Apparently, individual genius lies in the organization of the brain as much as its size.

ANATOLE FRANCE	2 lbs 4 ozs
WALT WHITMAN	2 lbs 15 ozs
DANIEL WEBSTER	4 lbs
OTTO VON BISMARCK	4 lbs
OLIVER CROMWELL	5 lbs
LORD BYRON	5 lbs
VICTOR HUGO	5 lbs

toire of behavior.

But going back to averages, why are mammals in general brainier than reptiles in general? Why were the first mammals brainier than their dinosaur contemporaries? This seems a puzzle, because the mammals and the dinosaurs both arose out of the same reptilian stock. The answer is probably connected with the nocturnal lifestyle of the mammal, and with the fact that it used the senses of smell and hearing for survival, rather than the sense of sight.

For the dinosaur, active during the day, sight was the most important sense, and the response to a visual cue was immediate. The scene taken in by the dinosaur's eye told its brain nearly everything the brain had to know immediately, and without the need for reflection and analysis. See a small, moving object; eat it! See a large, moving object; flee!

The brain was small in the early dinosaurs, and it remained small throughout their 140 million-year history. During that enormously long period their bodies improved and diversified in many ways, but their brains grew very little. Apparently, a small brain, with wired, automatic responses to a visual stimulus, sufficed for the dinosaur's survival. But the mammal, forced into a nocturnal habitat by the competition with the reptiles, could not be guided by vision during its nightly forays; it perceived the world around it through smells and sounds. In that circumstance lay the seeds of new growth in the mammal's brain.

Smells and sounds: they are very different from visual images. A smell, for example, does not depict the object itself; it only gives a hint or trace of its presence. Perhaps the prey was here some time ago and then departed, leaving a trail of scent; it must be tracked with patience and a skill born of experience; or the scent may have been left by a reptilian predator; then the mammal must remember the reptile's habits, and plan the night's activities accordingly; or it may be the elusive perfume of the female in heat. What are her habits? Where did she go?

Memory, planning, and a wisdom born of experience are critical for survival in the shadowy world of smells. The reflexive, nonthinking reactions of the dinosaur — See! Act! — will not do; such immediate and direct responses to olfactory cues are rarely useful. A small brain has room only for simple circuits, commanding automatic responses. Large brains, with space available for thought, for the analysis of subtle clues, for storing the memory of past experiences, and for planning future actions, are essential to the animal that relies on smells.

Life in a world of smells places extra demands on the size of the brain in still another way. A visual image — a glance at a scene in the forest — brings a wealth of information directly to the eye. Innumerable details, in all shades of light and dark, are imprinted on the retina; everything is there, available for immediate action. But a smell contains no details; a smell is just a single thing — one particular mixture of molecules that strikes the chemical receptors in the nose.* A few molecules of one kind can mean a tasty grub; a few molecules of another kind can signify a mate nearby; and a third kind may recall an entire region in the forest, some hunting ground familiar from past forays. Every detail of that forest region would be conjured up by this one odor — the bog, the rotting stump with its rich harvest of grubs, the reptile's lair. We humans still have this capacity, inherited from our mammal ancestors who rooted about in the dark of the night 100 million years ago. A smell can bring to mind an emotion, or a person not seen for years, or an entire period of one's life.

*One of the mysteries of smell is the fact that all the smell-sensitive cells in the nose are identical, yet they send different messages to the brain for different kinds of molecules. How do the smell cells do this? No one knows.

Where is the wealth of information stored, that a smell conjures up? Not in the molecule that enters the nose; a molecule does not have a picture engraved on its surface of a bog, or a stump, or a lair. A molecule is a simple thing — only a little cluster of atoms, like carbon and oxygen; it contains no pictures whatsoever. The picture is in the animal's memory, waiting to be brought to life when the nose signals a familiar odor; then it springs magically into the conscious mind. An animal dependent on the interpretation of smells for survival must have a large brain, with a capacious memory in which the experiences of an entire lifetime are stored, like a book ready to be opened to any page on command from the nose.

A keen sense of hearing was also important to animals that moved about in the dark of night. The brain of a nocturnal animal must have additional circuits for interpreting sounds, assessing their direction, and comparing them with information provided by other senses. This circumstance expanded the centers of the brain connected with hearing, and contributed further to the growth of the mammal's brain.

Now we can see why the brains of the early mammals became superior to the brains of the dinosaurs. Because the mammals were active at night, they were forced, like a blind person, to memorize a map of their surroundings, with odors as their guide to the map; and, again like a blind person, they could not act impetuously; they had to plan their actions. Thus, in each generation the mammals with larger and better brains, and greater powers of memory and planning, were more likely to survive the perils of the time and leave offspring. Throughout the years in which the dinosaurs reigned, natural selection worked steadily to prune the stock of the mammals, reducing the numbers of the small-brained and less intelligent, and augmenting the numbers of the large-brained. Over the course of time, the average brain size in the early mammals gradually increased. The first stage in the evolution of intelligence was completed.

Large brains, improved body design, warmblooded metabolism and parental care — these were advantageous traits for the early mammals.* Apparently they were not advantageous enough, for the mammals never overthrew their reptilian overlords. In spite of their many advantages, they remained subordinate to the reptiles for 100 million years.

But then, starting about 80 million years ago, the dinosaurs, like the mammal-like reptiles before them, began to decline in numbers, and the Age of the Reptiles slowly drew to a close. Toward the end, they went more rapidly; in one fossil-rich region in the western part of the United States, the rocks contain the bones of roughly 80 varieties of dinosaurs; while the rocks in another region, only a few hundred thousand years later in geologic time, contain the remains of only half as many.

Sixty-five million years ago, the last of these splendid creatures became extinct. They left no direct heirs, and no close relatives other than the birds, who share with them a common ancestry in those lively, rooster-sized reptiles that used their teeth and talons to such good advantage in the forests of a vanished world.

Why did the dinosaurs, so wondrously designed for their time, and so successful for millions of years, disappear from the earth? Many explanations have been advanced. Some scientists believe the mammals stole the dinosaurs' eggs and ate them. But the mammals had been around for millions of years; why should their consumption of dinosaur eggs suddenly become so overwhelming as to doom every member of this

*Warmbloodedness and brain size are connected. Brains use up a great deal of chemical and electrical energy; they burn ten times more energy per pound than the rest of the body. An animal cannot have a large brain unless it also has a warmblooded metabolism, to provide the brain with an abundant and continuous supply of energy.

populous and varied group of reptiles? Other scientists look to a global catastrophe for the explanation — an explosion of a nearby star, spraying the earth with lethal radiation; or the devastating impact of a large asteroid as the cause. The trouble with these catastrophist theories is that the reptiles did not go out like a light, but declined gradually in number and variety over some 20 million years.

A less dramatic explanation proposes that the disappearance of the dinosaurs is connected with their small brains and lack of adaptability to changing conditions. The dinosaurs were as smart as any reptile should be, but that still left them with a rather low grade of intelligence. About 80 million years ago,

EARLY GROWTH OF THE BRAIN. These drawings compare the relative sizes of the regions devoted to smell (shaded area) in the brains of a reptile (*left*) and a small, insect-eating mammal (*right*) similar to the mammals that lived in the time of the dinosaurs. The mammal's greatly enlarged smell brain is the result of the nocturnal lifestyle of this animal.

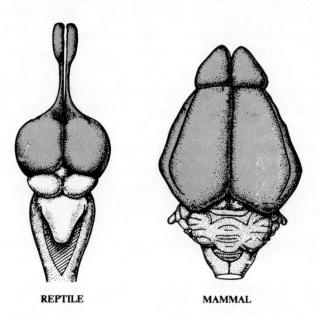

REPTILE   MAMMAL

the climate of the earth began to change. The world's weather had been mild and equable for more than 100 million years, throughout the reign of the reptiles, but now it took a turn toward cooler and drier conditions. The level of the oceans also dropped, and the continents of the world, which had been largely covered with water during the earlier period of warmth and moisture, now became exposed. At the same time, the continents had been drifting steadily away from the equator; when the dinosaurs first evolved, they existed mainly in the tropics and subtropics, where the weather rarely varied throughout the year, but now, because of the drift of the continents, many found themselves in northern and southern temperate zones, exposed to the cycle of the seasons.

Cool, dry weather and large areas of exposed land were not good for a reptile; they meant a change from the mild climate of an island warmed by the surrounding waters to the Siberian harshness of a continental interior, with its bitter winters and blazing summers. The poleward drift of the continents exacerbated these seasonal extremes of temperature.

An intelligent and small animal can cope with such stresses, because intelligence provides the flexibility needed to devise new strategies of behavior, such as hibernation, and small size makes the new strategies practical. In winter, the small animal digs a burrow and passes the season in relative comfort; in summer, many shady nooks are available to it.

The dinosaurs had neither of the essential traits required for survival; few were small, and none were intelligent. As a result, they were sorely pressed. The mammals, on the other hand, had all the important traits: they had small bodies which were easily tucked away; they possessed coats of fur and a warmblooded metabolism to help them further in surviving the extremes of hot and cold; and, most important of all, they had large brains, which endowed them with the priceless assets of intelligence and flexibility. In that changing world, flexibility must have been the most valuable trait an animal could have. Although the details may never be known, it seems likely that the secret of the dinosaur's extinction and the mammal's survival lies in this circumstance.

THE GROWTH OF THE HUMAN CEREBRAL CORTEX OUT OF THE SMELL BRAIN. The highest realms of thought of which the human brain is capable reside in the wrinkled outer layer of the brain known as the cerebral cortex. This region of the human brain grew out of an area devoted to the analysis of smells, in the brain of the little forest mammal that lived 100 million years ago.

The brain of the elephant shrew (*below*) probably resembles the brain of those primitive mammals that lived in the time of the dinosarus. Because the early mammals were active at night, smell was more important to them than vision; they navigated by smell, and smells provided the locations of food, mates and enemies. Gradually a thin coating of grey matter grew over and covered the part of the mammal's brain devoted to analyzing smells. In this coat of grey matter the mental activity took place by which the little animal coordinated smells with its other senses, and arrived at a decision or a plan of action. The coating of grey matter, called the neopallium or "new cloak", is the shaded area in the illustrations *below* and on the page *opposite.*

Apparently this kind of mental activity was of great value, because the neopallium continued to grow in the later generations of mammals. In the course of 100 million years it evolved into the massive cerebral cortex of the human brain (*shaded area, opposite*).

In humans, sensations of smell are the only signals that pass directly to the cerebral cortex. All other sensations pass first to a reception center called the thalamus for preliminary previewing. This circumstance harks back to the days when the cerebral cortex was evolving out of the centers of smell in the brains of our forest ancestors. The direct connection running from the nose to the human cerebral cortex explains the fact that an aroma can evoke extraordinarily vivid memories of past events.

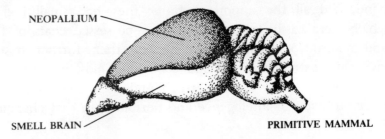

NEOPALLIUM

SMELL BRAIN

PRIMITIVE MAMMAL

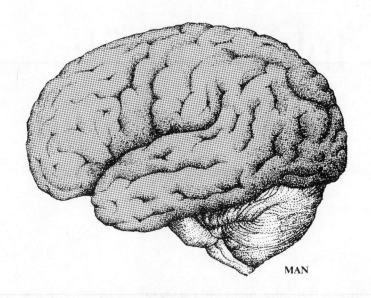

MAN

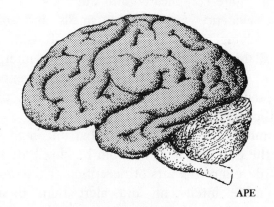

APE

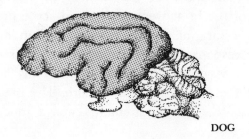

DOG

4·Into the Light: the Vision Brain

During the long night of the reptiles, the mammal's brain had become quite complicated and quite good at interpreting the subtle evidence provided by smells and sounds. The dinosaurs, relying mainly on their sense of sight, had gotten by on small brains, with a simple wiring for interpreting the evidence of their eyes. In their world, everyone was dim-witted — predator and prey alike — and simple, direct-action brains sufficed for the needs of all.

The mammals no longer lived in that uncomplicated world, in which the only threat came from hulking brutes of low intelligence, devoid of the arts of deception. These animals were generally more intelligent and alert than their reptilian forebears had been. In the society of the mammals, the strategies of the hunt were subtle, and acute perceptions were required for survival. A simple brain for handling vision, with automatic responses to everything it saw, was no longer adequate. The vision brain of the reptile was a good starting point, but the mammals needed more. The brain that evolved in response to their requirements was a much better brain than any that had existed before. The appearance of this new brain in the mammals marked the second great advance in the evolution of intelligence.

Picture a little mammal as it crosses the forest floor 50 or 60 million years ago. It lives in a world of mammals; all are clever; some are its enemies. The mammal moves past a tree, and its brain registers the form of the trunk — a columnar shape — and the rough texture of the bark. It perceives the dappled pattern of light and dark in the leaves of the branches above. Is a predator lurking behind the trunk? Is a leopard concealed in the leafy canopy overhead?

As the mammal skirts the tree, its eye perceives it from different angles and different patterns of light fall on its retina. The object *looks* different from each angle; it must *be* different. How does the brain of the mammal know this object is still the tree it saw a moment ago? It is not easy to synthesize the impressions of an object seen from different angles, at different distances, in different levels of light, and recognize that they are the same *thing* — an object with a form and substance independent of the eye of the observer.

The mammal's brain works the trick by using a very large memory, and fancy computing circuits that allow it to learn by experience. The mammal on the forest floor has seen many trees in its lifetime; its memory has tucked away impressions of all kinds of trees; it has learned how the appearance of a tree changes with angle, light and distance. Drawing on this stored knowledge, the mammal's brain quickly, without conscious thought, calculates how the appearance of this tree and its leafy canopy should change from moment to moment as the eye's line of sight changes. The brain compares its calculations with the unfolding evidence of the eye; if the brain's calculation agrees with the evidence received from the eye, the brain flashes a reassuring signal to the mind: "This object is the same safe tree that appeared on the eye's retina a few moments before. Relax; continue feeding," says the brain. Or it flashes an alarm: "The object, viewed from the new angle, does not agree with the predicted form. It is a threat. A predator is concealed in those branches; take flight!"

The computing problem presented to the brain is quite difficult. Its decisions cannot be precise; they require common sense and a judicious tolerance of error. If the brain insists on precise agreement from moment to moment between its calculations and the changing appearance of the tree, every breeze that rustles the leaves creates a false alarm. Such a brain, crying "Wolf" all the time, is a burden rather than an asset; the possessor of that brain, its search for food interrupted continually, grows emaciated. But if the brain's tolerance for error is too large, it never detects the predator in the tree. The possessor of this brain is also doomed.

Such judgments are far more complicated than the stereotyped responses of the reptile. They require a brain with additional circuits for thinking and learning by experience, and with many additional cells of grey matter for storing the remembrances of past experiences. These judgments also require an eye that will furnish the brain with detailed information, so that it can make the fine discriminations upon which the life of the animal depends. Yet the mammal's eye cannot send the mammal's brain every feature of the picture it sees. When the mammal looks at a tree, its retina is imprinted with an image containing millions of items of detail. Every other tree produces more millions of visual details, all different from those in the first image. No brain, not even the human brain, could accommodate so much information. In some way, the mammal's brain extracts the essence of "tree" from this mass of detail and stores that essence in its memory. How did the forest mammal's brain do this? How does the brain of modern man create its perceptions of reality out of the flickering lights that dance across the screen of the retina?

THE PROBLEM OF PERCEPTION. The eye takes in this chaotic patchwork of light and shadow and passes it on to the brain. After a moment, the brain sees the patchwork for what it is — a dalmatian.

The human brain accomplishes this feat because it has a large memory, with a capacity for storing many visual images from the past; it has seen and stored away the shapes of many dogs, and perhaps many dalmatians. The brain also has circuits that set up a comparison between the stored images of dogs and the image received from the eye. When these circuits indicate close correspondence, a recognition signal flashes into the conscious mind. How close does the correspondence have to be to trigger the signal? The answer is different for every person. Some see the shape of the dalmatian at once; others never do. No two brains have identical circuitry.

The answer is very surprising. Brains work like computers; the brain receives a picture from the eye in the form of electrical signals, which pass to the rear of the brain and into the region known as the visual cortex. In the visual cortex, nerve cells are connected together to form circuits, which resemble the circuits in a computer. These circuits test the signals from the eye for meaningful patterns and match them against old patterns stored in the brain's memory. If the patterns match, other circuits send signals to the higher centers of the brain saying, "We know what this is; now tell us what it means."

Scientists have invested a great deal of effort in finding out precisely how these circuits in the brain work. Their objectives are larger than the understanding of vision; they are really interested in finding out how the entire brain works. How does the human brain think? How does it learn? How does it remember? Their enterprise is a bold one. The human brain is more complicated than the astronomer's universe; it is the most complicated object science has ever tried to understand. Nonetheless, extraordinary progress has been made in unraveling the circuits of the brain, and the way these circuits work.

The story starts with experiments on a simple brain – the brain of the frog. This brain is similar to the brain of the reptile, but about 25 million years older in its ancestry and somewhat more primitive. Like the reptile, the frog relies to some extent on smell and touch, but depends mainly on sight for its survival. A frog lives and dies by the evidence of its eyes; it relies on vision to reveal the darting insect, the hawk circling overhead, or the snake moving through the grass. The eye of the frog is connected to its brain by an optic nerve, which carries messages back to the brain from the light-sensitive cells in the retina. In some ways the retina is like the film at the back of a camera. However, the rays of light falling on the retina do not make a permanent record of the scene, as they do on a piece of film. Instead, the light falling on the cells of the retina generates electrical pulses that travel along the optic nerve and enter the brain. These electrical signals vary from moment to moment, as the scene changes or the eye looks in different directions.

It might be expected that the brain would receive a detailed picture of the scene, which would be displayed somewhere and examined by the mind, in the same way a person looks at a picture on a television screen. However, experiments have shown that the frog's brain never gets to see the detailed picture. Attached to the back of the frog's retina are certain nerve cells, or neurons, that work like computers. Each computing cell is connected to a small part of the retina, and sees only a part of the complete scene. The computing cell analyzes the part it can see, and then it sends on to the brain, not the picture on the retina, but only something about that picture which it thinks the frog's brain should know.

According to the experiments, the computing cells send signals to the frog's brain if one of the following four things happens: a moving object enters the frog's field of vision; or, a moving object enters the field of vision *and stops there*; or, the general level of illumination in the field of vision suddenly decreases, i.e., the sky has suddenly darkened; or, a small, rounded dark object enters the frog's field of vision, and moves around in an erratic fashion.

What is the meaning of the first signal — a moving object? In a frog's world, stationary objects such as rocks and branches can be ignored, but all moving objects are potentially dangerous. The first signal is a general alarm.

The second signal, which indicates that the moving object has entered the frog's field of vision and stopped, means the potential danger has become real; a predator is stalking the frog. The predator may be a heron, or a snake, or a small boy.

The third signal — a general darkening of the sky — signifies that the stalking predator has drawn near, and its shadow has fallen on the frog; or perhaps a hawk, or some other bird of large wingspread, undetected by the frog up to that point, is directly overhead. Destruction is imminent. The frog's brain responds to this set of electrical pulses by energizing motor nerves that run to the muscles of the leg — and the frog jumps to safety.

These three kinds of computing cells are used by the frog's brain as predator-detectors. What is the meaning of the fourth signal, a small, black spot that moves about erratically? Clearly, this object is a fly or some other kind of insect. When the moving-black-spot signal reaches the brain, the brain responds automatically with an electrical pulse that travels along the motor nerves to the tongue — and the fly is inside the frog.

The computing cells still send the "black-spot" message to the brain if the frog's eye sees the dark spot against a shifting background of light and dark patches, such as would be formed by grass and leaves rustling in the breeze. For the frog to receive the "bug-alert," it is only necessary that the spot be moving relative to the entire background.

Similar experiments on the European toad show that its brain is wired up as a "worm detector." The toad's brain responds to any long, thin, moving object — the essence of worm — that enters its field of vision. The brain sends electrical messages along the motor nerves to the body, and the toad executes an automatic sequence of actions: it turns towards the long, thin object, ejects its sticky tongue, catches the worm, gulps, and wipes its mouth.

The worm detector and the bug-detector seem very simple. What could be easier than to see a moving black spot? Actually, the bug-detector requires a considerable amount of computing machinery in the frog's eye and brain.* The starting

---

*There is no special significance in the fact that some of the computing is done by the frog's eye and some is done in its brain; all the "computing"

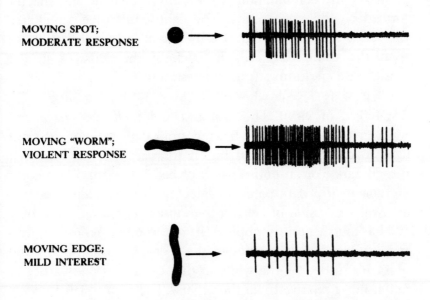

MOVING SPOT;
MODERATE RESPONSE

MOVING "WORM";
VIOLENT RESPONSE

MOVING EDGE;
MILD INTEREST

THE TOAD'S WORM DETECTOR. The electrical signals on the right were recorded when a small probe was placed in a 'computing' cell behind the toad's eye. They indicate the messages sent from the toad's eye to its brain as the eye caught sight of various kinds of objects. The signals are in the form of pulses. A rapid burst of pulses means an intense response.

In the illustration *above*, the arrows represent the direction of motion of a moving object. A moving spot elicits some interest, and the eye sends a moderatly intense signal to the brain (*top*). A line moving in the direction of its length triggers the "worm" response, and an intense burst of pulses goes to the brain (*middle*). A line moving perpendicular to its length is not wormlike, and a weak signal goes to the brain (*bottom*).

Toads and frogs can only see moving objects with extremely simple shapes. They never see a tree or the pattern on a butterfly's wings. These animals live in a visually impoverished world.

cells really belong to the brain. The eye is actually a part of the brain's surface that has grown out of it during the early part of the embryo's development. The eye remains connected to the brain by a stalk — the optic nerve — which is also a part of the brain. The eye and the optic nerve are dif: ferent from all other organs and nerves of the body in this respect.

point for the computation is the pattern of light and shadow formed on the frog's retina. The light-sensitive cells that line the retina see this pattern as a mosaic of light and dark spots; then the computing cells in back of the retina scan the mosaic and perform a series of tests on it.

What tests reveal whether a "bug" is moving across the frog's field of view? The first test is for *changes*: have any spots in the mosaic just changed from dark to light, or from light to dark? The second test is for *size*: has only one cell turned dark, or an entire patch of cells? The third test is for *movement*: if a dark patch is detected, is this patch sweeping across a succession of cells in the mosaic, one after the other?

The fourth test is applied if the computing cell reveals many moving dark patches in the scene, as when leaves and grass are rustled by the wind; in that case, does one particular patch move relative to all the others? The fifth test is for *stop-and-go motion*; does the moving dark patch stop and start, and stop again? If so, the strength of the signal is intensified; the computing cell has detected a bug.

These tests could be performed mathematically on an electronic computer. The procedure for carrying out the tests would include several thousand separate steps, and could be worked out by a computer programmer in three or four months. Such programming tasks are regarded as fairly difficult in the computer profession. Writing such a program requires a considerable amount of skill and experience. Who wrote the frog's program?

Darwin discovered the answer to that question: Nature wrote the frog's program. In generations long past, ancestral frogs with eyes and brains wired to respond to dark, rounded, erratically moving spots caught many flies. They were well-nourished, and lived to produce small frogs which tended to inherit this helpful trait; while frogs with a different wiring, that led them to ignore the image of the moving spot, or

A frog can starve in the midst of plenty (p. 70).

distracted them into the pursuit of other kinds of moving shapes, went hungry and produced fewer and less healthy offspring. In this way, the number of well-wired frogs increased, while the number of the poorly wired animals diminished, and their kind gradually disappeared from the population of frogs.

The changes in the nature of the frog were imperceptible from one generation to the next, but over the course of millions of generations they created an effective bug-catching program in the frog's brain.

It is interesting that although that brain was very small, even in proportion to the frog's small body, it was large enough to hold the circuits for the bug-catching program and the predator-detecting program as well. These programs only require a small brain because they are stripped to their bare essentials; the bug-catching program, for example, concentrates on the essence of "fly" — a small, dark, erratically moving spot — and shields the frog's limited brain from other distractions. Of course, the frog's program is stereotyped and unintelligent; it has no way of coping with challenges a frog never met before. Frogs placed in front of freshly killed, motionless insects will starve to death; according to the program in the frog's brain, an object that does not move is not an insect (page 69).

This is the trouble with small brains. They have no room for flexible responses to new challenges and opportunities. The brain of a frog or a reptile contains several hundred million neurons, all wired to carry out programs of instinctive behavior. The brain of a person also contains several hundred million neurons, wired up in the same fashion as the neurons in the reptile's brain; in fact, these neurons and their wiring are directly inherited from our reptile ancestors. But in man, on top of this wired, reptilian brain sits an additional mass of billions of neurons, largely blank and unwired at birth, and ready to absorb a lifetime of learning and experience.

5·Brains and Computers

Circuits, wires and computing are strange terms to use for a biological organ like the brain, made largely of water, and without electronic parts. Nonetheless, they are accurate terms because brains work in very much the same way as computers. Brains think; computers add and subtract; but both devices seem to work on the basis of the same fundamental steps in logical reasoning.

All arithmetic and mathematics can be broken down into these fundamental steps. Most kinds of thinking can also be broken down into such steps. Only the highest realms of creative activity seem to defy this analysis, but it is possible that even creative thinking could be broken down in this way, if the subconscious mind could be penetrated to examine the processes that appear at the conscious level as the flash of insight, or the stroke of genius.

The basic logical steps that underlie all mathematics and all reasoning are surprisingly simple. The most important ones are called AND and OR. AND is a code name for the reasoning that says, "If 'a' is true *and* 'b' is true, then 'c' is true." OR is a code name for the reasoning that says, "If 'a' is true *or* 'b' is true, then 'c' is true." These lines of reasoning are converted into electrical circuits by means of devices called "gates." In a computer the gates are made out of electronic parts — diodes or transistors. In the brain of an animal or a human, the gates are neurons or nerve cells. A gate — in a computer or in a brain — is an electrical pathway that opens up and allows electricity to pass through when certain conditions are satisfied. Normally, two wires go into one side of the gate, and another wire emerges from the other side of the gate. The two wires coming into the gate on one side represent the two ideas "a" *and* "b". The wire going out the other side of the gate represents the conclusion "c" based on these ideas. When a gate is wired up to be an AND gate, it works in such a way that if electrical signals flow into it from both the "a" *and* the "b" wires, an electrical signal then flows out the other side through the "c" wire. From an electrical point of view, this is the same as saying, "If 'a' *and* 'b' are true, then 'c' is true."

When the gate is wired as an OR gate, on the other hand, it permits electricity to pass through the outgoing, or "c", wire if an electrical signal comes into the other side through either the "a" wire *or* the "b" wire. Electrically, this is the same as saying, "If 'a' *or* 'b' is true, then 'c' is true."

How do these two kinds of gates do arithmetic? How do they carry on a line of reasoning? Suppose a computer is about to add "1" and "1" to make "2"; this means that inside the computer a gate has two wires coming into it on one side, representing "1" and "1", and a wire coming out on the other side, representing "2". If the gate is wired as an AND gate, then, when electrical signals come into it through both of the "1" wires, it sends a signal out the other side through the "2" wire. This gate has added "1" and "1" electrically to make "2".

Slightly different kinds of gates, but based on the same idea, can subtract, multiply and divide. Thousands of such gates, wired together in different combinations, can do income tax returns, algebra problems and higher mathematics. They can also be connected together to do the kinds of thinking and reasoning that enter into everyday life. Suppose, for example, that a company distributes several different lines of goods, and its management assigns a computer the task of keeping a continuous check on the inventories in these various product lines. Inside that computer, certain gates will be wired as AND gates to work in the following way: two wires coming into one side of the gate carry signals that indicate "stock depleted" and "sales volume heavy." If the stock is depleted *and* the sales are brisk, the gate opens, and a decision comes through: Order more goods!

OR gates are just as important in reasoning. Suppose that the same company also relies on its computer for guidance in setting prices. That means that a certain gate inside the computer is wired as an OR gate; coming into one side of this gate is a wire that indicates cash flow, another wire that indicates prices charged by a competitor for similar products, and a third wire that indicates the inventory in this particular product. If the company needs cash, *or* it is being undersold by its competitors, *or* it has an excess inventory, then the decision gate opens and a command comes through: Cut prices!

In a simple computer, the gates are wired together permanently, so that the computer can only do the same tasks over and over again. This kind of computer comes into the world wired to do one set of things, and can never depart from its fixed repertoire. A computer that solves the same problems in the same way, over and over again, is like a frog that can only snap at dark, moving spots; if either kind of brain is presented with a novel situation, it will react stupidly, or not react at all, because it lacks the wiring necessary for a new response to a new challenge. Such brains are unintelligent.

Larger, more complex computers have greater flexibility. In these computers, the connections between the gates can be changed, and they can be wired up to do different kinds of things at different times; their repertoire is variable. The instructions for connecting the gates to do each particular kind of problem are stored in the computer's memory banks. These instructions are called the computer's "program." When a computer expert wants his machine to stop one kind of task and start another, he inserts a new program into the computer's memory. The new program automatically erases the old one, takes command of the machine, and sets about doing its appointed task.

However, this computer is still not intelligent; it has no innate flexibility. The flexibility and intelligence reside in its programmer. But if the memory banks of the computer are extremely large a great advance in computer design becomes possible, that marks a highlight in the evolution of computers comparable to the first appearance of the mammals on the earth. A computer with a very large memory can store a set of instructions lengthy enough to permit it to learn by experience, just like an intelligent animal. Learning by experience requires a large memory and a very long set of instructions, i.e., a complicated program, because it is a much more elaborate way of solving problems than a stereotyped response would be. When a brain — electronic or animal — learns by experience, it goes through the following steps: first, it tries an approach; then it compares its result with the desired result, i.e., the goal; then, if it succeeds in achieving its goal, it sends an instruction to its memory to use the same approach next time; in the case of failure, it searches through its reasoning or computations to pinpoint the main source of error; finally, the brain adjusts the faulty part of its program to bring the result into line with its desires. Every time the same problem arises, the brain repeats the sequence and makes new adjustments to its program. A large computer has programs that work in just that fashion. Like a brain, it modifies its reasoning as its experience develops. In this way, the computer gradually improves its performance. It is learning.

A brain that can learn possesses the beginnings of intelligence. The requirements for this invaluable trait are, first, a good-sized memory, and, second, a wiring inside the brain that permits the circuits connecting the gates to be changed by the experience of life. In fact, in the best brains — judging brain quality entirely by intelligence — many circuits are unwired initially; that is, the animal is born with a large number of the gates in its brain more or less unconnected with one another. The gates become connected gradually, as the animal learns the best strategies for its survival. In man, the part of the brain filled with blank circuits at birth is greater than in any other animal; that is what is meant by the plasticity of human behavior.

Large computers have some essential attributes of an intelligent brain: they have large memories, and they have gates whose connections can be modified by experience. However, the thinking of these computers tends to be narrow. The richness of human thought depends to a considerable degree on the enormous number of wires, or nerve fibers coming into each gate in the human brain. A gate in a computer has two, or three, or at most four wires entering on one side, and one wire coming out the other side. In the brain of an animal, the gates may have thousands of wires entering one side, instead of two or three. In the human brain, a gate may have as many as 100,000 wires entering it. Each wire comes from another gate or nerve cell. This means that every gate in the human brain is connected to as many as 100,000 other gates in other parts of the brain. During the process of thinking innumerable gates open and close throughout the brain. When one of these gates "decides" to open, the decision is the result of a complicated assessment involving inputs from thousands of other gates. This circumstance explains much of the difference between human thinking and computer thinking.

Furthermore, the gates in the brains of an animal or a human do not work on an "all-or-nothing" basis. The AND gate in a computer, for example, will only open if *all* the wires coming into it carry electrical signals. If one wire entering a computer gate fails to carry a signal, the gate remains shut. If every one of the 100,000 pathways into a gate in a human brain had to transmit an electrical signal before that gate could open, the brain would be paralyzed. Instead, most gates in the brain work on the principle of ALMOST, rather than AND or OR. The ALMOST gate makes human thought so imprecise, but so powerful. Suppose that 50,000 wires enter one side of a gate in a human brain; if this were an AND gate in a computer, all 50,000 things would have to be true simultaneously before that gate opened and let a signal through. In real life, 50,000 things are rarely true at the same time, and any brain that waited for such a high degree of assurance before it acted would be an exceedingly slow brain. It would hardly ever reach a decision, and the possessor of a brain like that would not be likely to pass its genes on to the next generation.

Real brains work very differently. Wired largely out of ALMOST gates, they only require that, say, 10,000 or 15,000 things out of 50,000 shall be true about a situation before they act, or perhaps an even smaller number than that. As a consequence, they are inaccurate; they make mistakes sometimes; but they are very fast. In the struggle for survival, the value to the individual of the speed of such a brain more than offsets the disadvantages in its imprecision.

A NEURON OR GATE IN THE BRAIN. The photograph *opposite* shows a typical neuron or 'gate' in the brain. The neuron is the object enclosed within the circle at the center of the photograph. This neuron is a few thousandths of an inch in actual size. The solid arrows pointing toward the circle indicate channels converging on the entrance side of the 'gate', bearing incoming messages from nerve cells in other parts of the brain.

The open arrow pointing away from the circle toward the bottom of the photograph indicates the output wire or fiber, leaving the exit side of the 'gate'. The output fiber is called an axon. If a sufficient number of incoming messages arrives at the entrance to the gate, the gate opens and a signal passes out through the axon.

The point at which an incoming nerve fiber makes contact with the neuron is called a synapse. A synapse in the brain is similar to a diode or transistor in an electronic circuit. It allows an electric current to pass in only one direction – the direction that leads to the gate or neuron.

The drawing *below* shows an enlarged view of the body of another neuron with a somewhat different shape. The entire body of the neuron is encrusted with little round caps. These are the synapses – junction points at which nerve fibers make contact with this neuron, carrying messages from other cells.

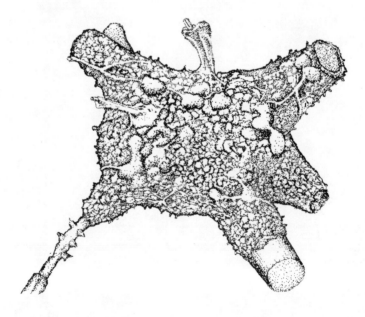

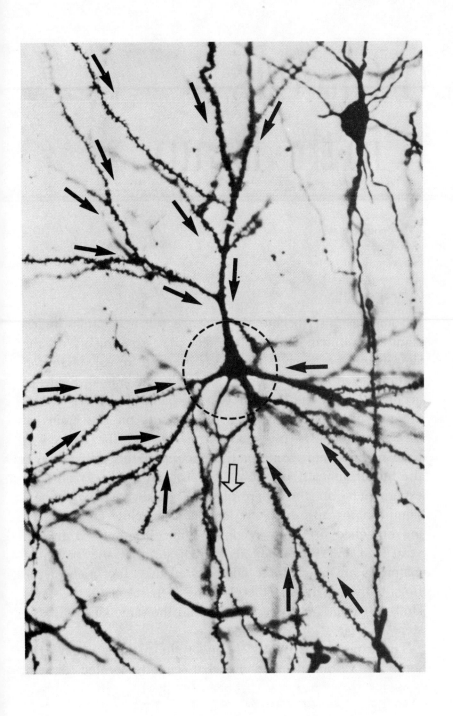

6·Circuits
in the Brain

Some scientists believe that science will never fully understand the human brain. They feel that experiments on the simple brains of animals like the frog cannot reveal valid information regarding the brain of an intelligent animal. Evidence accumulated in the last ten years suggests that these doubts are not justified. Remarkable experiments on the brains of several mammals — the rabbit, the cat and the monkey — have revealed how some of the circuits of the brain work in the higher animals, and how they probably work in man. The experiments on the brain of the monkey are particularly significant because monkeys and men are close cousins, having branched off from a common ancestor only 35 million years ago. All the experiments indicate that the brains of mammals are much more complicated than the brain of the frog, but, nonetheless, these brains seem to have the same kind of wiring, and to be built around the same kinds of basic steps in logic.

The experiments on the brain of the monkey involve the sense of vision. They show that circuits located directly behind the eye process the light signals from the retina in a preliminary way. This is the same as in the case of the frog. The circuits in back of the eye then send their results on to the brain. In the frog, the circuits behind the eye do most of the work, and very little is sent to the brain beyond 'insect' and 'predator' messages. In mammals, the reverse is true; the eye sends a wealth of information to the brain, where it is processed by complicated circuits in the visual cortex for use by the brain's higher centers.

The vision circuits in the brain of the mammal are vastly more ingenious than the circuits in back of the frog's eye. They prepare the brain for recognizing the difference between a Rembrandt and a Picasso, and the study of these circuits brings scientists one step closer to an understanding of the circuitry that underlies the brain's intelligent action.

Here is how it works. The light falling on the eye generates electrical signals in the cells of the retina. Nerve fibers, which are like wires, run from each cell of the retina to other cells immediately behind, which are wired to act as simple computers. Each computing cell combines the signals from several hundred separate cells in the retina, and forms a small, circular patch of light out of them. These cells are wired so that they will only send a signal to the brain when the patch of life they "see" consists of a dark spot silhouetted against a bright background, or the reverse.

This reduces the amount of detail in the image by a factor

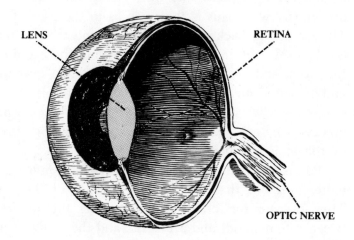

LENS   RETINA

OPTIC NERVE

THE HUMAN EYE. The lens of the eye focuses light onto the retina, which is lined by 100 million light-sensitive cells. The signals receive a preliminary processing in cells located behind the eye and then enter the optic nerve.

of about 100. As a result, the image sent on to the brain now looks like a high-quality photograph that has been broken up into small dots for printing in a newspaper or a book.

The million nerve fibers, or wires, that leave the eye are bundled into a cable called the optic nerve. The optic nerve runs from the back of the eye to an intermediate stop, or way station, buried deep within the brain.* From the way station, the wires carrying the image go to the visual cortex, located at the rear of the brain. The visual cortex is the brain's center of

*The way station, called the lateral geniculate nucleus, is located in the part of the brain called the thalamus. The largest number of nerve fibers from the eye goes to this way station. However, some fibers also connect to other parts of the brain that control eye movement, pupil dilation, and so on.

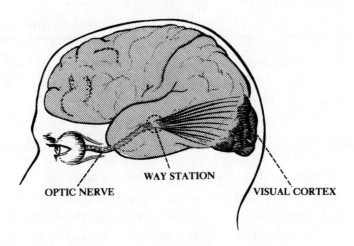

WAY STATION

OPTIC NERVE

VISUAL CORTEX

**PATHWAYS FROM THE EYE TO THE BRAIN. Electrical signals from
the eye pass through the optic nerve to a way station in the center of the brain,
and then to the visual cortex at the back of the brain.**

vision, and is devoted entirely to the reception and processing
of signals from the eye. The illustration shows the position of
the visual cortex in the human brain and the pathway that
connects it to the eye.

The visual cortex contains a detailed map of the image seen
by the eye. That is, every region on the surface of the retina
is connected to a corresponding region on the surface of the
visual cortex. If the eye sees a spot of light in one corner of a
scene, a patch of electrical activity will flash in a corresponding corner of the visual cortex. It is as if the eye were a television camera, and the visual cortex were a television set connected to the camera. However, a person cannot see the picture image displayed on his own visual cortex. The higher
centers of his brain get to see some aspects of the picture
eventually, but only after electrical circuits in the visual cortex

have worked on it extensively to bring out the most interesting features of the scene.

A great deal of computing goes on in the visual cortex. The computing is similar to what goes on behind the frog's eye, but much more elaborate. This computing, instead of being limited to the recognition of round, dark spots, can read the type in a page of print.

The part of the cortex that does the "computing" is a layer on the surface about a tenth of an inch thick. This computing layer is the grey matter of the brain, which is greyish in color in dead brains, but pink or lavender in living ones. Beneath the grey matter is a region about a twentieth of an inch thick, and more or less white in color, which contains the wires, or fibers, that run from one nerve cell to another. This layer of wiring is the white matter of the brain.

The arrangement is very much like that of the circuit boards in an electronic computer, in which the electronic components — diodes, transistors, and so on — which are equivalent to the grey matter in the brain — cover the top surface of the board, while the connections from one electronic part to another form a thick maze of wires — equivalent to the white matter in the brain — that runs beneath the board. Brain circuits and computer circuits are laid out on thin surfaces for the same reason; both must be spread out on a reasonably flat surface to provide easy access to the multitude of wires that connect one circuit with another.

Experiments on the brains of mammals have shown that the grey matter of the visual cortex contains circuits that do several distinct kinds of computing. In the simplest type of computation, a neuron in the visual cortex is wired so that it flashes a strong signal only if the part of the image it can see contains a sharp *line* or a *boundary* pointed in one particular direction, such as, for example, 30 degrees to the vertical. Furthermore, each neuron responds to a line that is located in one particular spot on the picture. These neurons are called simple neurons.

It is easy to figure out the kind of wiring that would make a simple neuron work. The signals entering the visual cortex from the eye represent the image as a pattern of circular spots. Suppose four wires, from four neighboring patches on the retina, are connected to one neuron. Suppose, to be specific, that the four neighboring patches happened to be lined up vertically in the retina. The neuron they are connected to will only fire — that is, transmit a signal — if electrical pulses come into it from all four wires at the same time.

This is the kind of circuit called an AND gate by computer experts, because the circuit will only send on a signal if it receives signals from "a" *and* "b" *and* "c" *and* "d". In this case, a, b, c and d are the four wires from the eye.

If the neuron the four neighboring patches are connected to fires a pulse, that means signals have come in on all four wires, which means, in turn, that at the moment, back in the retina, some part of the picture contained four dark spots stacked up vertically. The overlap of these spots makes a vertical line:

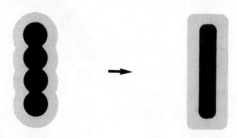

This is one kind of edge detector in the monkey's brain.

The same kind of circuit will also respond to lines pointed in other directions, depending on how the patches on the retina are lined up:

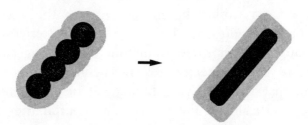

Simple neurons detect lines and boundaries. The wires from these neurons run to other neurons, called complex and hypercomplex neurons, which carry the computations a step farther. The complex neurons loosen up some of the restrictions on the response of the simple neurons; that is, a complex neuron responds to a line anywhere in its field of view, and not just in one spot.*

Hypercomplex neurons respond when their field of view contains two lines that meet to form an edge or a corner:

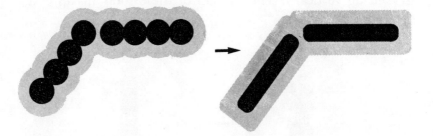

Hypercomplex neurons seem to terminate the brain's hierarchy of computations on the shapes of objects. Millions of con-

*The action of the complex neurons probably depends on the kind of wiring called the OR circuit.

fusing patches of light have been reduced to edges, boundaries, and corners. This is a beautiful scheme; edges, boundaries, and corners are an effective and economical way of capturing the essence of a picture without unwanted detail. The drawing of a cat below shows how it works — how an image can be put together with those simple elements.

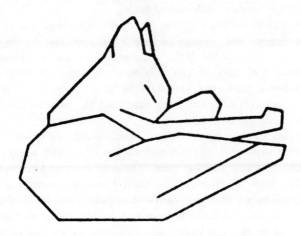

This is how the mammal sees things; this is the compromise evolution has worked out between seeing too much and seeing too little. Computing experts try to imitate nature with programs that teach a computer how to search out boundaries and corners in a picture. An artist uses the same technique when he deftly sketches a subject with a few expert lines. The artist does it consciously, with a skill born of experience, but the brain of every human performs the artist's trick without conscious effort, using the programs wired into the visual cortex.

HOW THE BRAIN SEES (*opposite*). This chart shows the electrical signals recorded by a probe inserted into a single cell in the visual cortex of a monkey. A black bar (left) was placed in the monkey's field of view before each record was obtained. At top, the black bar was oriented nearly vertically, and no signal resulted. In the next record, the bar was tilted at a greater angle, and produced a relatively small signal. The third record, which corresponded to a 45 degree tilt of the bar, produced an intense burst of activity in the brain cell. The last two records show that as the bar was tilted back toward the horizontal, the strength of the signal diminished again.

Experiments like these showed that the monkey's visual cortex contains cells that respond only to lines of specified orientation. As the probe entered the visual cortex of the brain, it passed through distinct layers of cells that responded to lines of gradually increasing tilt. The cells could discriminate between lines that were as little as 10 degrees apart. (Ten degrees is the angle between the hour hand and the minute hand of a clock when the time reads 12:02.)

Each small area in the visual cortex contains a column of nerve cells that looks at one small part of the image on the retina and searches this part of the image for edges and corners. The cells are wired so that they can compute the angle of any edge they find. They send a message about their conclusions into a higher part of the brain, where the bits and pieces of the scene are put together for examination by the conscious mind.

The illustration of the monkey's brain *below* shows the region in the visual cortex at the rear of the brain that responds to edges and angles of tilt in the image imprinted on the monkey's retina. It also shows other regions in the visual cortex that respond to color, distance, depth and movement. One of the remarkable discoveries in this field of brain research is the fact that each region contains a crude map of the image imprinted on the animal's retina, spread out across the surface of a small part of the brain. One region has a map of what the eye sees printed in color; another has a "distance map"; a third shows the depth in the scene, i.e., the way it would appear through a stereoscopic viewer; and the fourth shows moving objects in the scene.

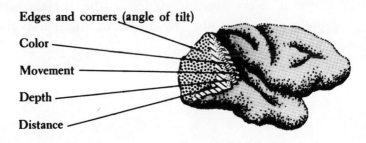

Edges and corners (angle of tilt)

Color

Movement

Depth

Distance

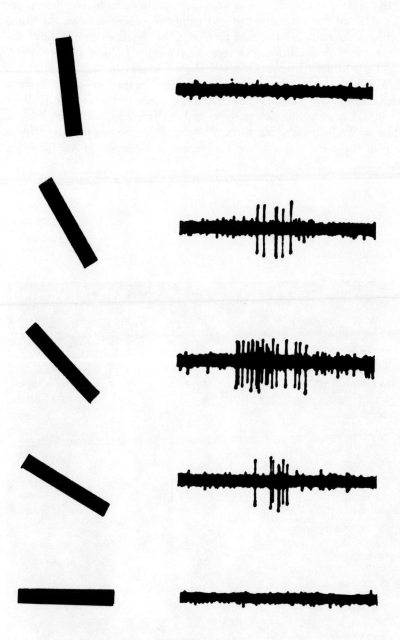

THE VISUAL CORTEX. The sequence of photographs on these pages was obtained from the scientists who did the experiments on the visual cortex in the monkey. The first picture in the sequence *(below)* shows the visual cortex in the brain of a monkey subject used in the experiments. The brain is seen from behind and slightly above the head. A slice has been removed from the cortex. The second picture *(opposite)* shows a cross-sectional view of the slice. The dark bands in the cross-section are the brain's grey matter, i.e., its gates and circuits. The light bands are the white matter, or connecting wires, that run from gate to gate. An enlarged view of the area in the dotted line rectangle is shown on the following page.

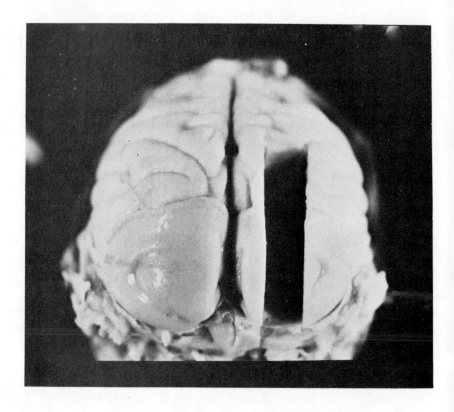

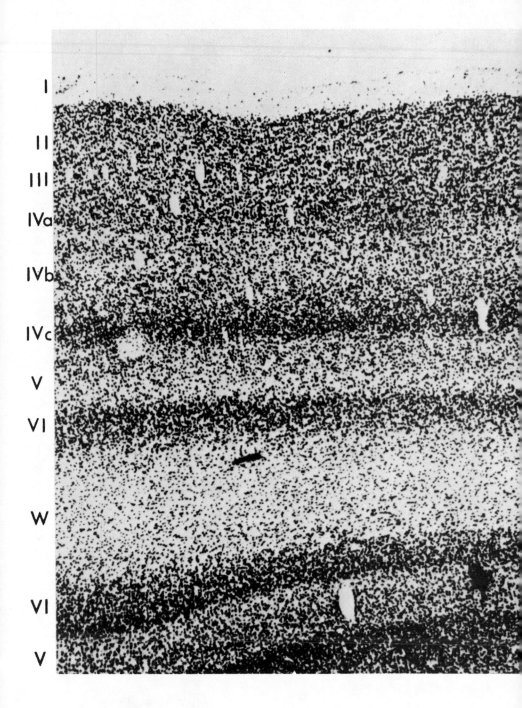

I

II

III

IVa

IVb

IVc

V

VI

W

VI

V

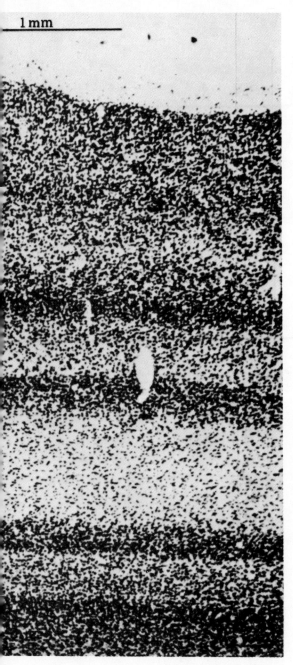

1 mm

MAGNIFIED VIEW OF THE VISUAL CORTEX. This photograph is an enlarged view of a small part of the visual cortex, about a quarter of an inch across, corresponding to the area in the dashed rectangle on page 91. Several layers of grey matter are visible. These are closely packed neurons that act like gates in a computer. The neurons in each layer seem to have different functions; "simple" cells are mostly in the middle layer, while "complex" cells and "hypercomplex" cells are mostly in the upper layers.

The light band near the bottom of the photograph is "white matter" containing the nerve fibers or connections running into and out of the visual cortex from the eye and other parts of the brain.

This minute section of brain tissue contains several million neurons or "gates". The surface of the brain as a whole is packed with 100 million neurons per square inch. The latest semiconductor circuit chips being developed by Japanese and American computer manufacturers contain nearly as many electronic components per square inch as the cortex of the brain.

7·A Guiding Hand

It is in the nature of biological evolution that it always proceeds slowly. A better eye and a better brain evolved in the mammals because they were advantageous traits, and those who possessed them in the highest degree were favored in the struggle for survival. But in any given generation the differences from one animal to another are slight. No mutants suddenly arise among the mammals, with sharper sight than ever existed before; no brain suddenly appears, with new and miraculously effective pathways for vision. These improvements in the design of eye and brain evolved gradually, through countless small changes in the separate parts of each organ.

The changes occur by accident in the disruptive passage of an x-ray or cosmic ray through the cell, or in the process by which the cell divides, or by the scrambling of the genes of the parents at conception. Many changes are deleterious to the organism, and they are eliminated gradually in later generations; but a few, by the purest accident, may be favorable, in the sense that they improve the design of the animal, even if in ever so slight a degree.

For a greatly improved eye or brain to appear suddenly, a thousand such changes must occur at once in a single animal, all accidental, and yet all in a favorable direction. That would be as unlikely as to toss a coin in the air and have "heads" come up a thousand times in a row.

In fact, it works quite differently. According to Darwin's theory of evolution by natural selection, the favorable change in eye or brain may occur in only one animal in a generation, or in none; but whenever it occurs it tends to be preserved, because it enhances the prospects of the individual for survival. Over the course of tens of thousands of generations, and a million years or more, the steady accumulation of innumerable slight improvements, each contributing to survival, will create a new animal out of the old.

The idea is simple, and there is no difficulty in seeing how it could explain the trunk of the elephant, or the neck of the giraffe. In each generation the giraffes with slightly longer necks than average reached higher into the foliage, and were better nourished and more likely to produce well-nourished offspring, who would inherit the favorable trait of a long neck. And so, by imperceptible degrees, from one generation to the next, the neck of the giraffe evolved to its present length.

These slow changes in the proportions of the body are accounted for in a straightforward manner by Darwin's theory. It is not so easy to accept that theory as the explanation of an extraordinary organ like the brain, or even the eye.

The eye, in particular, gave much trouble to Darwin in the development of his ideas on evolution, and much comfort to his opponents. The eye is a marvelous instrument, resembling a telescope of the highest quality, with a lens, an adjustable focus, a variable diaphragm for controlling the amount of light, and optical corrections for spherical and chromatic aberration. The eye appears to have been designed; no designer of telescopes could have done better. How could this marvelous instrument have evolved by chance, through a succession of random events? Many people in Darwin's day agreed with theologian William Paley, who commented, "There cannot be a design without a designer."

Darwin himself wrote, "To suppose that the eye, with all its inimitable contrivances . . . could have been formed by natural selection seems absurd in the highest degree." He continued to make changes in the discussion on the eye in later editions of *The Origin of Species,* always aware of the difficulty posed to his theory by such well-designed organs of the body. But his summation of the arguments for the evolution of the human eye is masterful:

> It is scarcely possible to avoid comparing the eye with a telescope. We know that this instrument has been perfected by the long-continued efforts of the highest human intellects; and we naturally infer that the eye has formed by an analogous process. . . . Further we must suppose that there is a power, represented by natural selection or the survival of the fittest, always watching each slight alteration in them, preserving each which in any way or in any degree, tends to produce a distincter image. Suppose each new state of the instrument to be multiplied by the million; each to be preserved until a better one is produced, the old ones to be all destroyed. . . . Variations will cause the slight alterations, will multiply them almost infinitely, natural selection will pick out with unerring skill each improvement. *Let this process go on for millions of years; during each year on millions of individuals. . . May we not believe that a living optical instrument might thus be formed superior to one of glass?* [italics added]

This passage captures the essence of Darwin's theory of evolution. The critical ingredients in evolution are time, many random variations, and "a power . . . always watching" and always working towards the improvement of the organ.

According to Darwin, time and innumerable chance variations can work miracles. But people have remained skeptical of that conclusion, from Darwin's time down to the present day. For one thing, there seems to be no direct proof that evolution can work these miracles. Darwin's theory suggests that a fish can change into a man, but no one has ever witnessed that miraculous transformation. The fossil record indicates that when the dinosaurs disappeared, the small mammals evolved into whales, elephants and many other kinds of creatures, but no one has ever seen a tiny animal metamorphose into a whale or an elephant. The evidence for these extraordinary transformations is indirect, and buried in the fossil record.

And then, there is the question of the creative power in evolution. According to Darwin, that power lies in nature; but what is nature? Darwin based his theory of evolution on artificial selection — the development of new strains of pigeons or dogs by animal breeders. There, man does the selecting, who does the selecting in evolution?

And finally, there is the role of chance. If Darwin was correct, man has arisen on the earth as the product of a succession of chance events occurring during the last four billion years. Can that be true? Is it possible that man, with his remarkable powers of intellect and spirit, has been formed from the dust of the earth by chance alone? It is hard to accept the evolution of the human eye as a product of chance; it is even harder to accept the evolution of human in-

MIRACULOUS TRANSFORMATION OF THE MAMMALS. The fossil record reveals that a small, ratlike animal — the basic mammal — evolved simultaneously into horses, pigs, whales, elephants and other modern mammals during a period of some 30 million years following the disappearance of the dinosaurs. The basic mammal moved into all the miches — land, sea and air — previously occupied by the reptiles, and became adapted, according to Darwin's theory, to the requirements of life in the new habitats by changes in its form. This remarkable phenomenon, which has occurred many times in the history of life, is called radiation by evolutionists, because all the lines of the many new kinds of animals radiate from one common source — in this case, the ancestral, furry little creature.

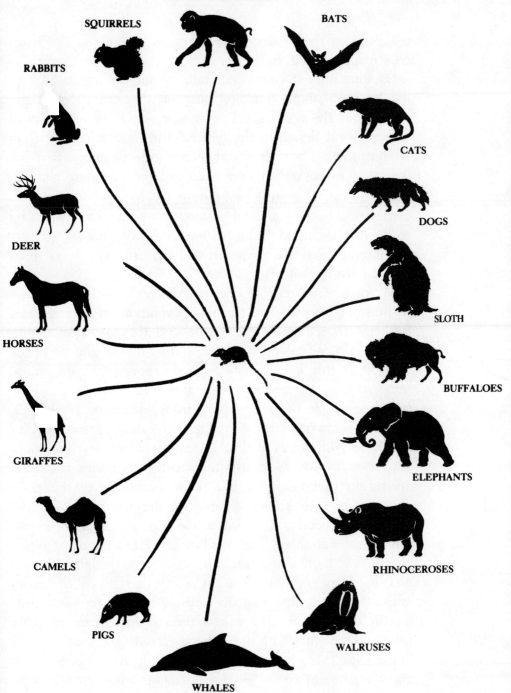

RABBITS

SQUIRRELS

MONKEYS AND APES

BATS

CATS

DOGS

SLOTH

DEER

HORSES

BUFFALOES

GIRAFFES

ELEPHANTS

CAMELS

RHINOCEROSES

PIGS

WALRUSES

WHALES

99

telligence as the product of random disruptions in the brain cells of our ancestors.

Darwin himself was uncertain about the matter. "My theology is a simple muddle," he wrote; "I cannot look at the universe as the result of blind chance, yet I see no evidence of beneficent design in the details." But most scientists today do not share Darwin's doubts; they are convinced that his theory of evolution removes the need for a guiding hand in the Universe. The great evolutionist George Gaylord Simpson expressed a nearly universal opinion among scientists when he wrote that evolution "achieves the aspect of purpose without the intervention of a purposer, and has produced a vast plan without the action of a planner."

My own views on this question remain agnostic, and close to those of Darwin. The theory of evolution seems complete; it seems to require no forces beyond the forces known to science. Yet, when you study the history of life, and step back to look at this long history with the perspective of several hundred million years, you see a flow and a direction in it — from the simple to the complex, from lower forms to higher, and always towards greater intelligence — and you wonder: Can this history of events leading to man, with its clear direction, yet be undirected? Scientists tend to feel that they know the answer to that question, but their confidence in the completeness of their knowledge may not be justified.

On the other hand, scientific faith in the proofs of evolution seems warranted. The fossil evidence in support of evolution is now fairly complete, and much more complete than it was in Darwin's day. This is particularly true of man's ancestors, of whom essentially nothing was known when Darwin lived. Discoveries by paleoanthropologists in recent years have filled in much of the missing record of human origins. Some gaps exist, but new finds are made nearly every year, and the moment seems to be approaching when the complete human lineage will be known. The continuity of the fossil record, even in its present form, proves that man did, in fact, arise out of simpler and less intelligent animals by a succession of minute improvements over hundreds of millions of years. Nearly every important bone in the human body can be traced back to the skeletons of the first fishes that left the water 350 million years ago. The modifications in the bodies of the backboned animals run in a clear line from the fishes to the amphibians, the reptiles, the mammals, and, finally, to man.

As with all historical evidence, the proof of man's animal origins is circumstantial, but its cumulative impact is overwhelming. The fact of evolution is not in doubt.

Whether this long process, culminating in man, is the expression of a plan or purpose in the Universe seems to me to be a question beyond the reach of human understanding, or at least beyond the reach of science. Scientists have an interesting story to tell about the flow of events leading from the creation to man, but, as with questions of beginning and end in the Universe, to these larger questions of plan and purpose, science has no answer.

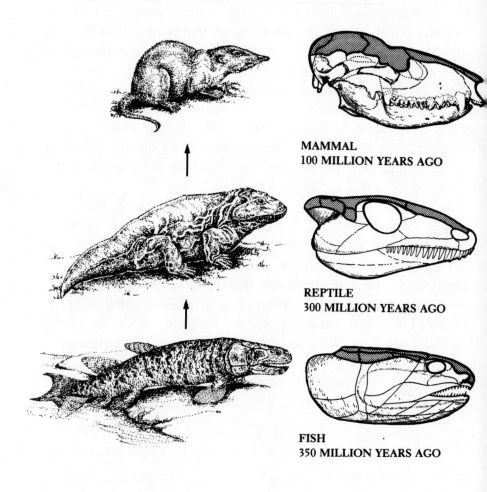

MAMMAL
100 MILLION YEARS AGO

REPTILE
300 MILLION YEARS AGO

FISH
350 MILLION YEARS AGO

FOSSIL EVIDENCE IN SUPPORT OF EVOLUTION. These illustrations show skulls and reconstructed shapes of several animals that were important links in the chain of evolution leading to man. The shaded area in each skull represents the same set of bones in the roof of the skull, shown as they gradually change through time. The bones in the roof of the skull grew larger as the size of the brain increased in our ancestors. Sequences of fossils like this form an important part of the evidence for man's evolution out of lower animals.

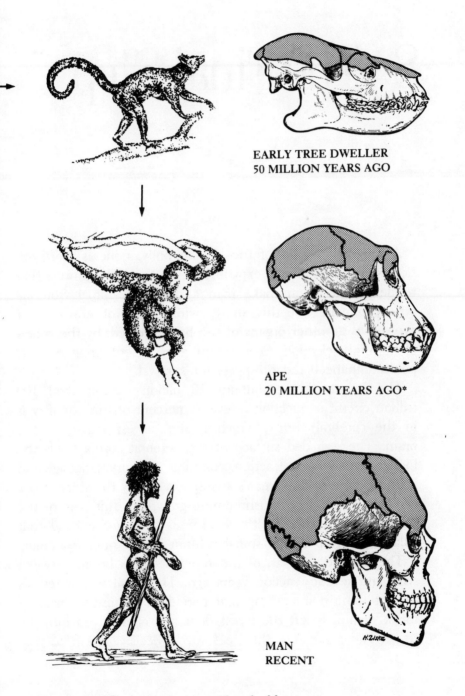

EARLY TREE DWELLER
50 MILLION YEARS AGO

APE
20 MILLION YEARS AGO*

MAN
RECENT

*Chimpanzee skull, representing ancestral stock of forest apes

8 · The Final Step

Among the organs of the human body, none is more difficult than the brain to explain by evolution. The powers that reside in the brain make man a different animal from all other animals. Yet this subtly wired mass of grey matter evolved as all other organs of the body evolved, by the action of natural selection in circumstances where greater brain power enhanced the prospects for survival.

The human brain contains 10 billion neurons and 100 trillion circuit connections. Seventy percent of this circuitry is in the cerebral cortex, which is the newest region of the brain. The wrinkled surface of the cerebral cortex holds the higher centers of abstraction, planning, memory, language and learning. The story of man's ancestry is largely the story of the cerebral cortex. The mushrooming growth of this part of the brain is responsible for the enormous size of the head that balances on top of the spinal column in the human animal.

The dramatic growth of the cerebral cortex began between sixty and seventy million years ago. The fossil record reveals that at some point during that period an inquisitive band of forest mammals left the forest floor and climbed up into the trees. The move into the trees was a decisive event for the evolution of the brain. No one knows the precise circumstances in which it occurred; why did one band of little mammals invade the trees, and not another? Why did one particular group of crossopterygians leave the water to invade the land, 350 million years ago? We do not know the answer to either question. We only know that a few animals migrated, others followed, and nature set to work to mold the forms of the migrants to the pressures of the new environment.

This is the way many evolutionary advances probably occur. An excess of curiosity, or some other behavioral novelty, draws a small number of animals into a new area of the environment. At first, only one animal ventures into the unfamiliar territory, or at most a few. These are the vanguard. If the strange environment offers advantages, others follow. Now natural selection begins to act on the small band of venturesome individuals, and adapts their bodies to the requirements of life in the new habitat.

Perhaps one mammal went up into the lower branches 70 million years ago because it was brighter than average, or more curious, or it may have possessed some variation in bodily traits — a sharper claw, a greater strength of limb, or a finer sense of balance — that enabled it to climb a little better. The new world of the trees offered special rewards to the climbing animal in exotic kinds of insects, luscious fruits, and greater protection from predators. Its brothers were bound to notice its success, and imitate its behavior. Among the imitators, some were better than others at climbing, or keeping their balance, or judging the distance from branch to branch. These individuals remained in the trees, and flourished there. They were favored for life in the new environment by the accidents of birth. The litters of the favored individuals were numerous and healthy; their numbers multiplied rapidly. The offspring tended to inherit the advantageous traits that fitted their parents for an arboreal life. Over many generations, a new strain of tree-adapted animals evolved.

Those first tree dwellers were the primates — ancestors of the monkey and the ape.* They are called primates from the Latin for "first," because their descendants also include man, who is first among living creatures.

The early primates carried good brains with them, inherited from their brainy mammal forebears. Life in the trees was difficult as well as rewarding, and good brains became better. The brain had been growing slowly but steadily since the advent of the mammals 200 million years ago. Now, with the move of their descendants to the trees, the brain began to grow faster.

Vision was a key factor in the accelerated growth of the tree dweller's brain. Vision had been important to the fishes and the reptiles, but in the early mammals it had been eclipsed by smell. Now, in the tree dwellers, it was of prime value again.

Scampering from branch to branch, the tree dweller must make accurate judgments of distance. A poor judgment can be fatal. A study of the skeletons of the gibbons, or tree apes, reveals that although these apes are the most accomplished acrobats in the primate world, one third fracture their limbs in serious falls at some point in their lifetimes.

The judgment of distance requires eyes that point straight ahead, with overlapping fields of view. Among the forebears of the primates, who had lived on the forest floor, the eyes had been located more to the sides of the head than the front, because stereoscopic vision was less important than the constant surveillance of the horizon in every direction for pro-

---

*A group of animals known as the prosimians also belong to the primates. The prosimians include the lemur of Madagascar, the tarsier of Borneo, and others. They are descended from an early stage in the evolution of the tree dweller, before that animal had evolved into the monkey.

tection against predators. But the panoramic view had disadvantages. Imagine a forest scene perceived by eyes that look straight out to either side, and lack the perception of distance. Each eye sees a flat, two-dimensional world, like the backdrop to a stage set. All animals seem to be equally far away, and the motionless, camouflaged bodies of predator or prey blend perfectly into the pattern of light and shadow. Now let the eyes move slowly forward, to a position at the front of the skull. Magically the set takes on depth, and trees and animals stand out from the background according to their true distance.

Stereoscopic vision was essential to the primate. How did his eye adapt to this need? The answer lies in the slight variations in bodily traits that occur from one individual to another in a population. Among humans, for example, some have close-set eyes, while in others the eyes are more widely spaced. Just so, among the early tree dwellers some had eyes that were closer to the front of the head than others. In these animals the fields of view of the eyes overlapped to a greater-than-average degree, and the power of stereoscopic vision was better developed than in their kin. Such individuals were less likely to suffer a serious fall from the trees, and more likely to survive to maturity and leave offspring who would inherit the advantageous trait of forward-looking eyes.*

The fossil record reveals the changes that occurred in response to these pressures. During the course of some millions of years, the eye sockets in the skulls of the early primates gradually moved around to the front of the head, until, in the monkey and the ape, they came to face directly

---

*Other animals have adapted to a tree-dwelling life differently. The tree squirrel, for example, lacks the stereoscopic vision and grasping hands of the monkey, yet it lives an acrobatic life in the forest canopy. The primates, with their well-developed eyes and hands, happened to represent the particular kind of adaptation to a tree-dwelling life that would later lead to the human line.

forward. This movement of the eye sockets took 20 million years.

As the position of the eyes changed, the brain began to change also. Eyes with overlapping fields of vision cannot calculate a distance by themselves; they must be backed up by a brain with range-finder circuits wired into it. Recent experiments on the monkey have turned up evidence for these range-finder circuits in the brain. The experiments are similar to the ones that revealed the presence of circuits in the monkey's brain for classifying shapes in terms of edges, boundaries and corners. In the range-finder experiments, as in the earlier experiments, a fine electric probe was inserted into the brain of the anesthetized monkey, and its penetration into the brain was increased slowly and carefully, so that it passed through the individual nerve cells in the visual cortex, one at a time. If one of these nerve cells fired while the probe was in contact with it, the tiny pulse of electricity emitted by the cell would be picked up by the probe and recorded on a sensitive instrument. If the experimenters found, for example, that one particular cell emitted a signal when the animal's right eye was open, but not when only the left eye was open, they would know that this cell was wired to the right eye.

EVOLUTION OF THE EYES AND HANDS IN THE TREE-DWELL-ERS (*opposite*). In the early tree-dwellers (*top*), the eyes have started to move to the front of the head for distance judgement. The long nose of this animal betrays its recent life in the world of smells on the forest floor. The front paws still have claws. They are not yet hands, but the digits have some dexterity and can spread and curl around a branch.

In a later stage in the evolution of the tree-dweller, represented by the tarsier (*middle*), the eyes face full forward, providing better stereoscopic vision and distance judgement. The nose has collapsed, indicating the reduced importance of smell relative to vision in the tree-dweller's life. The paw is now nearly a hand. Natural selection has trimmed the claws back to nails to permit full use of the fingers (*bottom*). This animal does not scratch its way up a tree, as a cat does; it grasps the branches like a monkey. Oversized pads on the fingertips strengthen their grip.

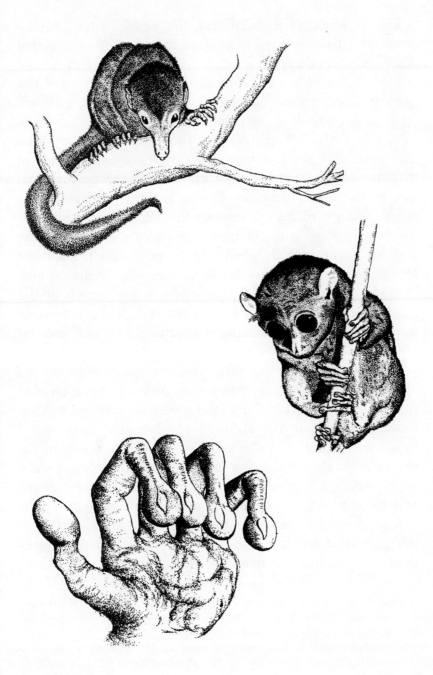

The experiments revealed that the visual cortex contains some cells that are wired to one eye only, and other cells that are wired to both eyes.

These latter cells are the rangefinders. They are wired like gates in a computer. The wiring is such that the gate will open and transmit a message only if it receives identical signals from the left eye and the right eye. This means that the muscles of the eye have swiveled the eyeballs about so that the left eye and the right eye are focused on the same spot in the scene. It is easy for the brain to arrange this; it sends messages to the eye muscles to stretch or contract until the double image in the brain merges into a single image. The position of the eyeballs when that happens, and the amount of tension or stretch in each eye muscle, will depend on how close the object is. If we could calibrate the stretch of the muscles in the eye, we could read the distance to the object from the length of these muscles as if they were the scale on a range finder.

The circuits that can "read" the eye muscles are in the visual cortex of the brain. Presumably these circuits evolved in the visual cortex gradually, as the eye sockets moved to their forward position in the skull. At any rate, the fossil record shows an expansion of the visual cortex in the brains of the early tree dwellers during the period in which the eye sockets were moving. This expansion contributed to the general enlargement of the brain in the monkey and the ape.

A heightened sense of color probably contributed further to the growth of the tree-dweller's brain. Color vision is highly developed among the monkeys, the apes and man.* Color

---

*Apart from the primates, few mammals possess a sense of color today, although many reptiles and birds have it, and the reptilian forebears of the early mammals probably possessed it also. But the early mammals lost their color vision; probably it disappeared in a general deterioration of the visual sense among the early mammals, as they descended into the long night of the dinosaurs.

heightens the value of vision; it lends the ability to spot fruit at the end of a branch, or distinguish friend from foe. In the world of the trees, color can be a matter of life and death.

As a sensitivity to color developed, new circuits for analyzing colors must have been added to the brain of the tree dweller, as well as circuits for combining color with other information. Color tells the brain one thing; shapes tell it another; the combination of the two reveals more than either sense would alone. But such associations of form and color do not spring out of nowhere; they are created in the brain, in new circuits and gates, all adding to the brain's volume. And there is the additional strain on the brain's memory, which now must include color, and the association of color with other qualities, when it stores away its recollections of an important event. More cells of grey matter are needed for the new storage bins. It is no wonder that the skulls of the tree dweller began to bulge outward under these pressures.

At this point, a third factor came into play, that worked with the new senses of color and stereoscopic vision to create another spurt of brain growth in our ancestors. The new development involved the transformation of the paw into a hand. An animal with clawed paws can climb, as the cat does, but much of the pleasure of an arboreal life is denied to it. It cannot divide its weight among the branches and grasp a tasty morsel at the limit of the leafy canopy; and it cannot leap from branch to branch, or tree to tree. An active life in the trees requires a well-shaped hand, with fingers that can curl around a branch; it also requires a different construction of the shoulder and hip, to permit the movement of the limbs in several directions, and a wrist that can rotate 180 degrees. No dog or cat can rotate the wrist, or raise its forelimbs above its head; only an ape or a human can do that, and, to some extent, a monkey. The fossil record reveals that these traits of hand and limb also appeared among the primates, in the course of 30 million years of change.

The animals that possessed the new bodily traits were very different creatures than their four-footed cousins who had remained below. The new creatures were, in fact, the monkey and his cousin, the ape.

The shape of the man can be seen in the monkey. When an animal possesses stereoscopic vision and a hand with supple digits, it can caress an object with its eyes, and feel the texture, form, weight and temperature with its hands. A new world of sight and touch opens before it — a rich, three-dimensional world of tactile experience. All the new sensations of sight and touch must be relayed to the brain of the monkey, and there they must be put together, digested and analyzed before they pass into the memory. New circuits were required in the brain of the monkey and the ape for that synthesis. Here was a development with an explosive impact on the evolution of intelligence.

The rest of the history of human evolution is almost a detail. The monkey did not change very much from the time of his appearance, 30 million years ago, to the present day. His story was complete. But the evolution of the ape continued. He grew large and heavy, and descended from the trees to take up a new existence on the forest floor, returning to the habitat his ancestors had forsaken millions of years earlier. Some additional millions of years went by, and a flourishing band of apes — ancestors of the chimpanzee and the gorilla — came to inhabit the forests of East Africa. Fifteen million years ago, a change in the climate of the earth deflected the course of evolution. It had happened before. Three hundred and fifty million years ago, a long drought set in on the earth, and the fishes left the water to create a new line of land animals; and 80 million years ago, another change in the earth's climate signaled the demise of the dinosaurs and

ROTATION OF THE LIMBS IN THE TREE DWELLERS. New kinds of joints were needed in the monkey and the ape. Man, descended from the same ancestors, is the only other animal that can raise an arm above the head.

opened the world to the mammals. The latest change in climate was a trend to cooler and drier weather that caused patches of open grassland to appear in the East African forest. As the dry weather continued, the grassland expanded, and the terrain was transformed gradually into the familiar savanna of modern Africa.

No one knows what happened next, because a gap of several million years appears in the fossil record at this point. But the forest apes were brainy and curious animals, more so than any before them. Perhaps the boldest and most inquisitive among them left the security of the forest to explore the strange environment of the open grassland. Apparently, some forest apes found the open land to their liking and remained there, because when the record picks up again, it shows the descendants of the apes well established on the savanna. But the progeny of the forest ape were no longer apes; they were a new kind of animal, who walked on two legs, with a manlike stride. And the new animal had a superior brain; relative to body weight, his brain was nearly twice as large as the brain of the ape.

According to the fossil record, this two-legged, intelligent animal appeared in Africa about four million years ago. His name is Australopithecus afarensis. He was a skillful hunter who competed with the other carnivores of his time, and held his own against the lion and the giant hyena. Australopithecus was a slightly built animal, only four feet tall, and lacking talons, slashing canines, or other natural weapons; but he survived unprotected. His weapon was his brain.

This brain was not large in absolute size; it was a third the size of a human brain, and weighed scarcely a pound; but, on the other hand, the animal it governed was also not large. Australopithecus, with a small body, required relatively few brain cells for the control of his muscles, and had more grey matter available for memory, planning, and abstract thinking.

Australopithecus achieved this brainpower two or three million years ago, several million years after his ancestors first moved onto the savanna. He changed somewhat in the next million years or so, evolving successively into the animal known as Australopithecus africanus, and then into the larger-bodied Australopithecus robustus. One million years ago, he became extinct.

But Australopithecus did not vanish without issue; a million years or so before he disappeared, he gave birth to another intelligent animal on the African savanna. The new creature is named Homo erectus — the man who stands erect. His stance was upright, like that of Australopithecus, but nobler; his body and countenance were more manlike and less apelike; and his brain was larger.

At first, the difference in brain size was not great, but the intelligence of Australopithecus remained nearly unchanged, while the brain of Homo continued to grow. The final step in the evolution of the brain had commenced.

Now the story has carried us to a time, about one million years ago, when Australopithecus was about to vanish from the savanna. Homo erectus already hunted with a formidable skill; no game was too large for his prowess, and his stone-tool technology was impressive. Suddenly, the brain of this animal began to grow at an extraordinary rate. No one knows exactly why that happened, but the making and use of tools is thought to be an important factor. The tool-making abilities of early man derived from his bodily traits, and ultimately from his erect posture, which had freed the forelimbs for other uses than knuckle-walking. During 60 million years of evolution, the paw had been transformed into a hand of extraordinary construction. The hand of Homo erectus was both supple and strong; it had the power grip that man uses when his fingers curl around a hammer or a club; and it had the precision grip that places thumb and forefinger in delicate opposition for picking up a small object or threading a needle. Homo also had eyes with overlapping fields of vision, and he had a brain with precise range-finding circuits and well-developed centers for the coordination of sight and touch. These were magical attributes; they could make a spear or an axe; but Homo also needed a center in his brain that possessed powers of abstraction, to see the club concealed in the knobby outlines of a stout branch, or the cleaver lurking in the rounded surface of a rock. An imaginary world must be created in the brain of the toolmaker, in which scenes are enacted that have never occurred — scenes in which one hard stone is struck against another to create a sharp edge; or the pelt of a dead animal warms the shivering body; or the darkness of the cave retreats before a burning torch.

Who among the early men had this kind of imagination? Only a few; as with modern man, the individuals in that time must have varied in their powers of abstraction and the quality of their imagination. Those who possessed these mental traits in the highest degree were the great inventors and theorists of their day. And then, as now, there were those with practical skills — the gifts of resourcefulness and ingenuity needed to put a theoretical idea to good use. These men, working with the theorists, created cunning strategies for the hunt, and effective weapons for felling game; they brought home the bacon. They and their offspring flourished; their genes spread through the population, and their traits were intensified among the early men.

Some watched the early heroes at their work, and grasped the principles of the new technologies sooner than others. They had quick minds. And a good memory was important also, for storing the technical wisdom of past generations.

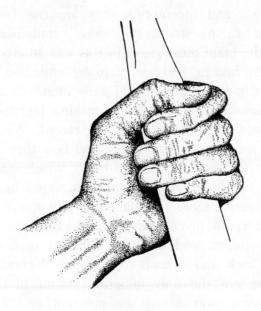

THE POWER GRIP

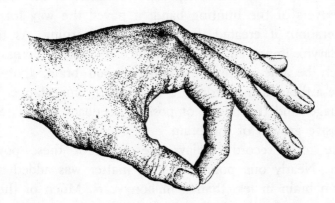

THE PRECISION GRIP

117

A good memory, a quick mind, powers of abstraction, resourcefulness and ingenuity — they are the faces of intelligence, and their seat is in the brain. If traits like these are intensified, the brain must grow. In this way an erect posture, leading to the making and use of tools, tended to accelerate brain growth in the population of early men.

Many relics of early man's tool-making technology have been preserved in the archaeological record. We know his tool-making skills at different times, and how they evolved in tandem with the evolution of the brain. When skillful hands made tools, the importance of tools increased in daily life; when tools became more important, the importance of a good brain, which could invent even better tools, also increased. Skill fosters wisdom; wisdom fosters skill. The process is called positive feedback, and it leads to exponential growth.

Tools were not the only important factor in the brain's growth. Language was at least as important, and also worked on the brain in positive feedback. When the human larynx began to evolve, along with other bodily parts needed for the framing of clear sounds, the importance of speech grew among the early men. Speech coordinated the complex maneuvers of the hunting band; it paved the way for social cooperation; it created the tradition of learning. As tongue and larynx improved, the importance of speech increased; as speech became more important, a good brain, that could frame a thought in words, also increased in importance. Tools, language, and the power of positive feedback stimulated the explosive growth of the brain.

The fossil record displays the result of these powerful forces. Nearly one pound of grey matter was added to the human brain in less than a million years. Much of the new growth was in the forepart of the brain's cerebral cortex, which is the center of the highest realms of creativity and abstract thinking. As the additional neurons crowded into the cerebral cortex of early man, his forehead bulged upward and outward, the face lost its brutish cast, and he began to assume the high-domed look of the intellectual. Homo sapiens — the Man of Wisdom — had arrived on the planet.

STONE TOOLS: THE FIRST HUMAN TECHNOLOGY. The chipped stone *above* is a so-called pebble tool, made by early man two million years ago. The stone *opposite* is a cleaver, with handle at bottom, made about one million years ago. These tools, found in the Olduvai Gorge in East Africa, are among the earliest man-made tools known. They were manufactured by the human ancestor known as Homo habilis, considered by most paleo-anthropologists as the first true man. Homo habilis lived about two million years ago and was the ancestor of Homo erectus.

Choppers and axes like these were indispensable in skinning and quartering large animals. They were made from very hard stones, usually pieces of congealed lava from nearby volcanoes, which had been worn to a rounded shape by tumbling in stream beds. Two or three hard blows on the stone with another "pebble", used as a hammerstone, created a chopper with a sharp cutting edge along one side. The rounded shape of the chopper fitted nicely into the palm of the hand. The axe was made with a dozen blows of the hammerstone. The thin chips or flakes that came off could be used as knives.

In spite of their simplicity, choppers and cleavers were effective butchering implements. Many are found in ancient campsites, scattered among the bones of large animals like the elephant that had been brought down by the

hunting prowess of early man. These tools, and the big-game hunting that accompanied them, are very impressive in view of the fact that the average brain size of Homo habilis, according to remains found with the tools, was about 650 cubic centimeters or 40 cubic inches–less than half the size of modern man's brain.

At that early stage in human evolution, the pace of technological progress was still very slow. Although the cleaver was made a million years after the chopper, the difference between the two tools in level of technology is mainly a matter of 2 or 3 blows of the hammer stone in making the chopper, compared to a dozen blows in making the cleaver.

EVOLUTION OF THE HANDS AND FEET. The different varieties of living apes have hands and feet that reflect how much time they spend on the ground. The comparison of hands and feet in these animals tells the story of the evolution of an erect posture in man's ancestors. The development of this upright posture freed the hands for carrying infants and food, and for making and using tools. Walking on two legs was a major factor in the subsequent evolution of the brain.

In the gibbon (*below*), which probably resembles the early tree apes, there is hardly any difference between hands and feet. However, the gibbon's digits are beginning to develop the suppleness that is an important characteristic of the human hand. In effect, the gibbon has four 'hands' — well-formed grasping appendages.

The gorilla spends a great deal of time on the ground, and has a foot and hand (*opposite*), that are quite different from each other. The foot is beginning to approach its human appearance, but the gorilla's big toe is still angled out to the side. The forward push off the ball of the foot, which marks the human stride, is not possible. The gorilla's hand is capable of neither a power grip nor a precision grip. The thumb is small, and the fingers are stubby.

In man — and in our ancestors as far back as Australopithecus afarensis, four million years ago — all five toes point forward. These feet have lost their grasping power and can no longer be used as hands, but they make powerful platforms for running and walking. The hand itself is fully formed with well-developed fingers and a large thumb, suitable for holding a hammer or turning a lathe.

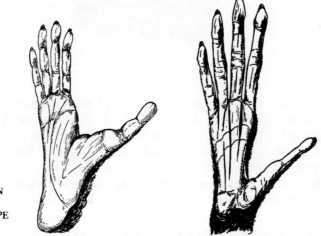

GIBBON
or
TREE APE

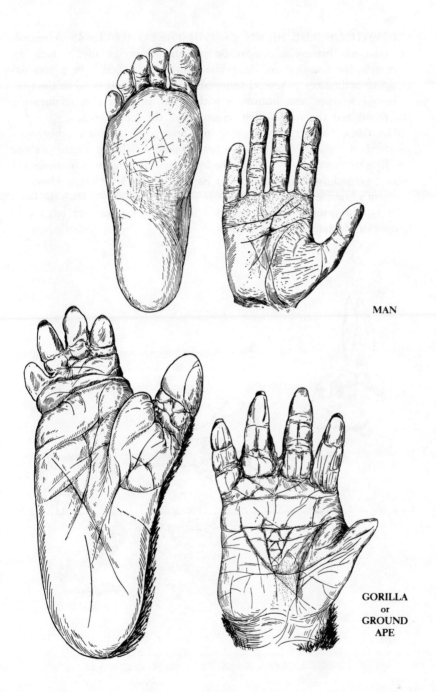

MAN

GORILLA
or
GROUND
APE

GROWTH OF THE BRAIN IN THE TREE DWELLERS. Mammals are relatively brainy in comparison to other forms of life. Among the mammals, the brainiest are the primates or tree dwellers. These drawings compare brain sizes in several mammals whose ancestors lived in the trees — monkeys, apes and humans — with brain sizes in two mammals — the rabbit and the wolf — that remained on the forest floor.

The rabbit has a relatively small and primitive brain. If a rabbit were enlarged to the size of a man, its brain would weigh one ounce — about one-fiftieth the size of a human brain. The wolf is unusually intelligent among ground-dwelling animals and has a relatively large brain. However, the wolf's brain is two times smaller relative to body weight than the brain of a typical primate — the monkey — and ten times smaller relative to body weight than the brain of man.

RABBIT

WOLF

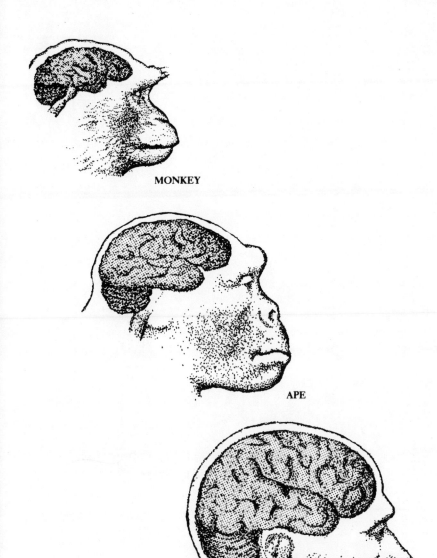

MONKEY

APE

MAN

9·The Old Brain and the New

The human brain consists of several different regions that evolved at different times. As each new section of the brain grew in our ancestors, nature generally did not discard the old parts; instead, they were retained and the latest section was built on top of them. Today the cerebral cortex, newest and most important region in the human brain, folds around and smothers the older and more primitive regions. Yet these regions have not been completely overpowered. They remain underneath, no longer in undisputed command of the body, but still active. These primitive parts of the human brain continue to operate in accordance with a stereotyped and unthinking set of programs that go back to the mammals on the forest floor, and back farther still to the brutish reptiles who spawned the mammal tribe. Experiments have shown that much of the human repertoire of behavior originates in deeply buried regions of the brain that once directed the business of life for our ancestors.

The brains of our reptile forebears were divided cleanly into three compartments: a front compartment for smell, a middle compartment for vision, and a rear compartment for balance and coordination. All three compartments grew out of the brain stem, a still more ancient collection of neurons at the top of the spinal column.

These arrangements were inherited from the simple brain of the fishes. The receptors for vision and smell were coordinated in a region between the smell brain and the vision brain, which was a command post called the diencephalon.* Here, the inputs from the different senses were compared and put together for a program of action. The basic instincts of survival — sexual desire, the search for food and the aggres-

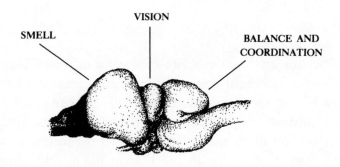

SMELL

VISION

BALANCE AND
COORDINATION

THE THREE-COMPARTMENT BRAIN OF THE REPTILE. This brain contains compartments for smell, vision, and balance. It is a simple brain, wired from birth with the programs or patterns of behavior needed for the reptile's survival. Remembering and learning play little role in the life of a reptile. Newly hatched snapping turtles, for example, find their way to the water by instinct, and at once begin to catch insects and small fish without parental instruction.

*The diencephalon is also called the thalamencephalon, from the Latin meaning "the couch on which the brain rests." In more advanced brains the diencephalon became the site of the thalamus and the hypothalamus.

sion responses of "fight-or-flight" — were wired into this region of the reptile's brain. Responses to the sex object, food, or the dangerous predator were automatic and programmed; the cerebral cortex, with its circuits for weighing options and selecting a course of action, did not exist.

When the mammals evolved out of the reptiles, their brains began to change. First, they developed a new package of instincts, related to the reptilian instincts for sex and procreation, but modified for the special needs of a mammalian lifestyle. Chief among these was the instinct for parental care of the young. Here was a revolutionary advance over the behavior of reptile parents, for whom the newly hatched young provided a tasty snack if they could catch them. But the reptile young were prepared to fight for their lives; they came into the world with all the needed programs of action wired into their brains. These hatchlings were miniature adults from the moment of birth. In the population of the mammals, on the other hand, the young arrived in a helpless and vulnerable state, and parental affection was essential for their survival. In these circumstances the mammal that lacked an instinct for the care of its young left few descendants. In the course of many generations, the traits of the indifferent parents were pruned from the stock of the mammals, and every mammal that remained was an attentive parent, and descended from a long line of attentive parents.

The new instincts of the mammals for parental care did not replace the older reptilian instincts; they augmented them. The ancient programs of the reptile brain — the search for food, the pursuit of a mate, and flight from the predator — were still essential to survival. As a result, the command post in the brain that controlled instinctive behavior grew larger. Its responsibilities now included parental care, in addition to its other burdens.

The brains of the mammals changed in another important

way, that was related to their nocturnal lifestyle. As these animals passed into their 100-million-year time of darkness, the vision brain diminished in importance and the smell brain expanded. Two bulbous swellings grew out of the smell brain, one on each side, packed with circuitry for comparing the input from the sense of smell with information yielded by the other senses. These swellings and their circuits had been present in the reptile's brain, but they did not dominate that brain. In the early mammal, for whom the sense of smell was more valuable than any other, the expanding globes of the smell brain gradually took over the functions of the main command post, and the older, reptilian centers of the brain diminished in importance.

The two swellings in the smell brain were the cerebral hemispheres. In the beginning, when smell was the main function of the cerebral hemispheres, these parts of the brain were modest in size and could be fitted into the cranium of the mammal without wrinkling or folding. Later, when the ruling reptiles disappeared and the mammals began to move about by day and rely on the sense of vision as well as smell, more circuits had to be added to the brain to receive the new information from the eyes and analyze it. The added circuits for vision were in the cerebral hemispheres, which swelled to an even larger size as a result.

A heightened sense of vision placed additional demands on the memory capacity of the mammal's brain. The circuits for this function also were located in the cerebral hemispheres. The hemispheres now grew at an even faster rate. Their surfaces, crammed into skulls of limited size, began to acquire the wrinkled appearance characteristic of a very brainy animal. The cerebral hemispheres also acquired a new name in the terminology of the brain. Being nearly all surface, these regions of the brain became known as the cerebral cortex, from the Latin word for the rind of a fruit.

The rate of growth of the cerebral hemispheres —now the cerebral cortex — was greatest in the monkey and the ape. The growth of the cerebral cortex accelerated further in man's immediate ancestors, and reached explosive proportions in the last million years of human history, culminating in the appearance of Homo sapiens.

The primitive region in the brain, that held the circuits for the instinctive behavior of the reptile and the old mammal, was now completely enveloped by and buried within the human cerebral cortex. Yet this ancient command post, relic of our distant past, is still active within us; it still vies with the cerebral cortex for control of the body, pitting the inherited programs of the old brain against the flexible responses of the new one.

Experiments suggest that parental feelings, source of some of the finest human emotions, still spring from these primitive, programmed areas of the brain that go back to the time of the old mammal, more than 100 million years ago. In one experiment, the cerebral cortex was removed from the brain of a female hamster, leaving only the reptile and old-mammal centers of instinctive behavior. Yet the hamster matured normally, showed an interest in male hamsters, gave birth to a litter, and was a good mother. In another experiment, when the cerebral hemispheres were left intact but the old-mammal centers of instinct were removed, the hamster lost all interest in its new-born young.

One part of the old brain, called the hypothalamus, is only the size of a walnut in the human brain, and yet a minute electrical stimulus applied to this region in the brain can create the emotional states of anger, anxiety or acute fear. The stimulation of nearby regions, only a few tenths of an inch away, produces sexual desire, or a craving for food or water.

The hypothalamus also appears to contain centers for aggression, killing, and fight-or-flight responses. A rat that will normally kill a mouse placed in its cage, no longer does so after these centers are removed. The action of the hypothalamus has a strong effect on personality. If a small electrical stimulus is applied to a particular part of the hypothalamus, an affectionate cat will turn into a biting, snarling animal, furious with the world; but its rage collapses instantly on cessation of the electrical current.

These experiments indicate that states of anger and aggression are created by electrical signals originating in the hypothalamus. The hypothalamus behaves as though it contains a gate that can open to let out a display of anger or bad temper. Normally, this gate is kept closed, but now and then the animal's senses tell its brain that its rights are endangered; a mate is lured away, food is stolen, or threat signals are received; and then the package of brain survival programs called "the emotions" comes into play, and an electrical signal to open the gate comes from some ancient center of instinct deep within the brain. Inhibiting signals from the cerebral cortex — the seat of reason — may quell the electrical disturbance in the old brain, and keep the gate closed; but if the perceived threat is very great, the electrical signals arriving at the gate are overwhelming, and the gate opens. And if these electrical signals are introduced artificially by the discharge of an electrode implanted in the brain, as in the experiments on the cat, the gate can open without provocation.

It is as if two mentalities resided in the same body. One mentality is ruled by emotional states that have evolved as a part of age-old programs for survival, and the seat of this mentality is in the old-mammal centers of the brain, beneath the cerebral cortex. The other mentality is ruled by reason, and resides in the cerebral cortex. A person who loses his temper sometimes becomes aware of the two mentalities within him; he feels he is outside himself, watching the display of anger, and wishing that it would cease, but power-less to end it. In these moments, the centers of reason in the cortex lose control over the primitive circuits buried in the brain, and stand by, watching, as these circuits take over the body. It is not without cause that people will speak of a person in such a case as having "lost his reason," or being "beside himself."

In man, the cerebral cortex, or new brain, is usually master over the old brain; its instructions can override the strongest instincts towards eating, procreation or flight from danger. But the reptile and the old mammal still lie within us; sometimes they work with the highest centers of the brain, and sometimes against them; and now and then, when there is competition between the two mentalities, and the discipline of reason momentarily weakens, they spring out and take command.

These properties of the human brain lead to a prediction regarding the life that will follow man. As nature built the new brain on top of the old in our ancestors, so too, in the next stage of evolution after man, we can expect that a still newer and greater brain will join the "old" cerebral cortex, to work in concert with the cerebral cortex in directing the behavior of a form of life as superior to man as he is to the ancient forest mammal.

THE OLD BRAIN AND THE NEW. The views of the human brain on the facing page show the brain stem, the most ancient part of the brain. The brain stem lies under the cerebral cortex and carries messages into and out of the brain from the remainder of the body. The brain stem goes back at least 400 million years and is a remnant of the first centralized nervous system in the history of life.

The cerebral cortex, which covers the brain stem, is the newest and largest region in the human brain and makes up 90 per cent of its mass. The cerebral cortex is the seat of memory, learning, and abstract thinking.

Lying under the smothering masses of the cerebral hemispheres, at the top of the brain stem, are the reptile brain and the brain of the old mammal (*below*), which contain the basic survival programs related to flight from danger, hunger, thirst, procreation and parental care. These parts of the human brain evolved between 100 million and 300 million years ago.

At the rear of the cerebral cortex is the cerebellum, or little brain, which coordinates the muscles of the body in complex maneuvers like waltzing or driving a car that involves the teamwork of dozens of muscles. When the body is learning a complicated sequence of movements, the brain has to think about each step in the sequence. After the sequence has been learned, the instructions to the body for executing it are stored in the cerebellum. They reside there permanently, like steps in a computer program, available for subsequent use without conscious thought. If, for example, the cerebral cortex decides to turn left at the next traffic light, it sends a command to the cerebellum, which takes over and executes the maneuver smoothly without further engaging the conscious mind. Of all parts of the brain, the cerebellum is closest to an automatic computer.

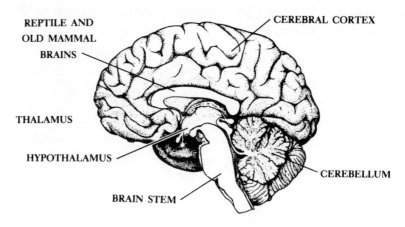

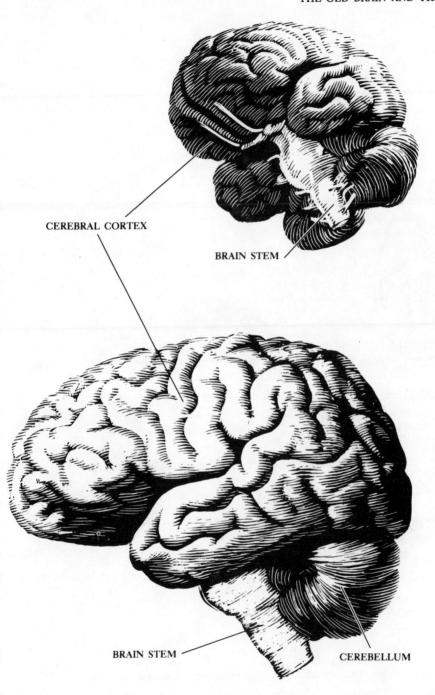

CEREBRAL CORTEX

BRAIN STEM

BRAIN STEM

CEREBELLUM

135

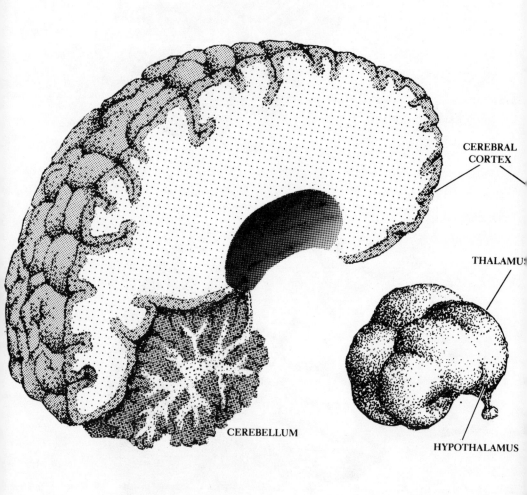

CEREBRAL
CORTEX

THALAMUS

HYPOTHALAMUS

CEREBELLUM

136

ONE BRAIN WITHIN ANOTHER. This exploded view of the human brain shows the cavity deep within the cerebral cortex that holds the reptile and old mammal parts of the brain. Much of the cavity is filled by the thalamus—twin masses of grey matter the size of robin's eggs. The thalamus acts as a reception center for messages from all senses except the sense of smell, sending some signals on to the higher centers in the cerebral cortex, and taking immediate actions on others. Messages from the nose pass directly to the cerebral cortex, in an arrangement that goes back to the early days of the mammals, more than 100 million years ago, when the cerebral hemispheres were first evolving out of the smell brains.

Under the thalamus lies the hypothalamus, a particularly important part of the old brain because it rouses the body and prepares it for actions appropriate to an emotional state. In times of stress, it is the hypothalamus that sends messages to the heart to quicken the pulse, and messages to the stomach to cease digestion and release valuable blood to the muscles for flight.

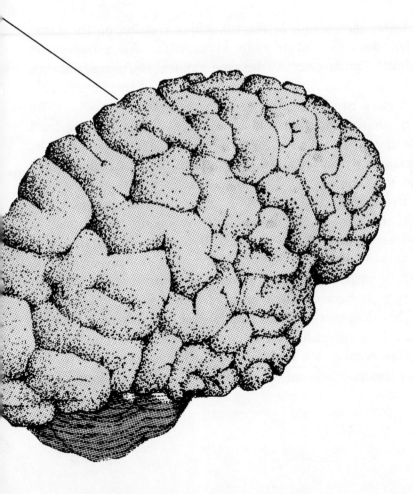

10 · Our Brain's Successor

Today man stands at the summit of creation on the earth. What does the future hold in store for this extraordinary animal? Perhaps he will become extinct, as Australopithecus did before him; more than 90 percent of all the forms of life that have existed on the earth have become extinct. Or he may survive unchanged into the distant future, a living fossil like the oyster. This fate may already be upon us, for the human body has changed very little in the past million years, and the human brain has not changed, at least in gross size, in the past 100,000 years. The organization of the brain may have improved in that period, but the amount of information and wiring that can be crammed into a cranium of fixed size is limited. The fact that the brain is no longer expanding, after a million years of explosive growth, suggests that the story of human evolution may be over.

This does not mean that the evolution of intelligence is over. It is reasonable to assume that human beings are not the last word in the evolution of intelligence on the earth, but only the rootstock out of which a new and higher form of life

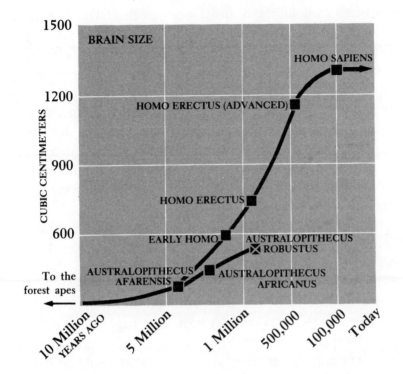

1500

BRAIN SIZE

HOMO SAPIENS

1200

HOMO ERECTUS (ADVANCED)

CUBIC CENTIMETERS

900

HOMO ERECTUS

600

EARLY HOMO
AUSTRALOPITHECUS
ROBUSTUS

To the
forest apes
AUSTRALOPITHECUS
AFARENSIS
AUSTRALOPITHECUS
AFRICANUS

10 Million YEARS AGO   5 Million   1 Million   500,000   100,000   Today

EVOLUTION OF THE HUMAN BRAIN. The fossil remains of man's recent ancestors show that the brain began to grow rapidly about one million years ago, after a long period of moderate growth in the monkey and ape line of evolution. About 250,000 years ago, the curve of growth began to level off, and in the last 50,000 years the size of the human brain has hardly changed. The human body has also changed very little in the last million years. Man is a nearly finished chapter in evolution.

According to the findings of Donald Johanson and Timothy White, the savanna ape, Australopithecus afarensis, gave rise to an australopithecine line as well as the human line. The australopithecines coexisted in Africa with early man until one million years ago, when they became extinct. The X on the chart marks the extinction of the australopithecine line. Their brain size at that time was about half the size of the brain of the early humans. Living in the same environmental niche with early man, they probably could not compete with his superior brainpower and toolmaking skills.

will evolve, to surpass our achievements as we have surpassed those of Australopithecus and Homo erectus. The history of life supports this conclusion, for it shows a seemingly inexorable trend toward greater intelligence in the higher animals. Apparently, among all traits of a living organism, none has greater survival value than the flexible, innovative response to changing conditions that we call intelligence. It seems unlikely that this trend in evolution, which has persisted for more than 100 million years, should suddenly stop at the particular level of mental achievement that we call "human." If the past is any guide to the future, mankind is destined to have a still more intelligent successor.

What form will our successor take? Judging by the history of man, the new form of life will resemble the old, but have a considerably larger brain. If this forecast is correct, the next species of intelligent life on the earth will be a creature like ourselves, but with a very large cranium and puny muscles.

Certain trends in modern technology suggest a very different vision of the future. Powerful forces of evolution are at work — cultural rather than biological — that could lead to a more exotic form of intelligent life, evolved out of man, but the child of his brain rather than his loins.

According to this vision, the new form of life is being created today in the laboratory of the computer scientist. It is an artificial life, made out of silicon chips rather than

BRAIN SIZE AND INTELLIGENCE. The brain of Neanderthal man *opposite* was as large as the brain of modern man in proportion to body weight, yet Neanderthal man became extinct 35,000 years ago and modern man survived. How can this fact be explained, if the ratio of brain weight to body weight is an indicator of intelligence?

The answer may lie in the fact that the Neanderthal brain has less grey matter in the forepart of the brain — called the frontal lobe — than the brain of modern man. Neanderthal man was highly intelligent, but the shape of his brain suggests that he was less gifted than modern man in the most creative realms of thought. For us, these realms signify music,

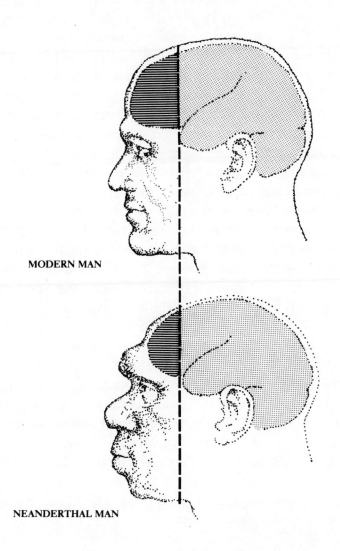

MODERN MAN

NEANDERTHAL MAN

art and science, but in early times they must have signified the possession of traits like innovativeness that had great practical value in the struggle against adversity. Average brain size is useful as a rough measure of intelligence in a population, but much also depends on the brain's internal organization.

141

neurons; yet it thinks, remembers, learns by experience and responds to stimuli. Its thinking is still simple; it is not very creative; but it is evolving at a lightning pace.

The suggestion seems absurd; how can the richness of human thought be compared to the mechanical thinking of a computer? It is true that the electronic brains of today are very primitive compared with the human brain; in fact, they have little going for them except a prodigious memory and some math skills. Yet the newest models can be wired up to follow an argument, ask pertinent questions, and write pleasing poetry and music. They can also carry on somewhat distracted conversations so convincingly that their human partners do not know they are talking to a machine.*

These are amiable qualities for the modern computer; it imitates life like an electronic monkey. As the computer gets more complex, the imitation gets better. Finally, the line between the original and the copy becomes blurred. In another 15 years or so — around 1995, according to current trends — we will see the computer as an emergent form of life, competitive with man.

Before we reject this vision, let us consider some recent developments in the computer industry in relation to the capacity of the human brain. At the present time, brains are greatly superior to computers. An average human brain weighs three pounds, consumes electrical energy at the rate of

---

*The following snippet of unrehearsed dialogue took place a few years ago between a computer and a human subject:
Subject: Men are all alike.
Computer: IN WHAT WAY
Subject: They're always bugging me about something or other.
Computer: CAN YOU THINK OF A SPECIFIC EXAMPLE
Subject: Well, my boyfriend made me come here.
Computer: YOUR BOYFRIEND MADE YOU COME HERE
Subject: He says I'm depressed much of the time.
Computer: I AM SORRY TO HEAR YOU ARE DEPRESSED

25 watts, and occupies a volume of one-tenth of a cubic foot. Within this small volume, the brain houses between 10 billion and 100 billion items of information. The first of the truly modern generations of computers is represented by the IBM 360 machine, which appeared in 1960. This computer has a memory that holds a few million items of immediately accessible information. Although the capacity of such a machine is far less than that of the brain, it consumes about 100,000 watts of electrical power, and occupies hundreds of cubic feet of space. Scaling the 360 computer upward in size, a machine matching the human brain in memory capacity would consume electrical energy at the rate of one billion watts — half the output of the Grand Coulee Dam — and occupy most of the space in the Empire State Building. Its cost would be in the neighborhood of $10 billion. The machine would be a prodigious artificial intelligence, but it would be only a clumsy imitation of the human brain.

The qualitative superiority of the brain over today's computers is even more striking than its compactness. Every cell, or gate, in the brain is directly connected to many other cells, in some cases to as many as 100,000. As a result, when we send a conscious impulse down to the recesses of the memory to summon forth a point of information, the cells in which this information is stored communicate on a subconscious level with thousands of other cells, and a wealth of associated images pours out at the conscious level of thought. The fruits of the subconscious activity are intuitive insight, flashes of perception and creative inspiration, all made possible by countless connections among the cells of the human brain.

The computer memory, in contrast, is like a set of pigeonholes stacked against a wall, with no thinking capacity in any pigeonhole, and no connections from one hole to another. Information can be placed in a pigeonhole or taken out of it, but there are no associations and thinking goes on elsewhere.

It is plain that the IBM 360 computer is a mini-brain in comparison with the product of millions of years of human evolution. On almost every count, it is hopelessly inadequate in comparison with the one-tenth of a cubic foot of grey matter that resides in the human cranium.

But the intelligent machine, still in its infancy, is evolving rapidly. The experience of the last three decades indicates that computer capability increases by a factor of ten every seven or eight years — a period that, in the jargon of the specialists, constitutes a computer generation. This seems a reliable rule of thumb for projecting the state of the computer art into the future. The record is clear. The first generation of computers, based on vacuum tubes, came into use in 1950. The second generation, which appeared about 1958, was based on transistors and was ten times faster. The third generation, represented by the IBM 360, was ten times faster still, and did its thinking with the aid of "chips," tiny squares of the metallic substance silicon. These silicon chips replaced both the transistor and the bulky vacuum tube.

The fourth generation of computers, based on better silicon chips, came into use in the 1970s. One of these machines, expanded to match the capacity of the brain, could be housed in a large office suite and would consume only 10,000 watts of electricity. It would still be clumsy, but almost worth building.

Further improvements, now under development in computer laboratories, will be ready for the market later in the 1980s. In the most advanced silicon chips thousands of electronic gates are crowded onto a surface with an area of a tenth of a square inch. By 1990, chips containing one million gates should be available. When that stage is reached, electronic circuitry will be nearly as compact as the circuitry of the brain. Million-gate chips will make it possible to build a computer of human capacity that can be placed in a trunk

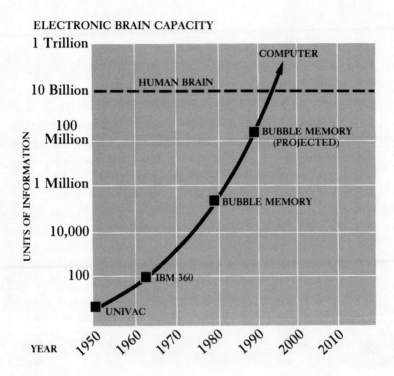

ELECTRONIC BRAIN CAPACITY

ELECTRONIC BRAIN CAPACITY. The compactness and power of computers have increased dramatically in the last few decades. Since 1950, there has been a millionfold increase in the amount of information that can be fitted into a computer memory with the same volume as the human brain. Trends in computer evolution indicate that a human-sized electronic computer will match the storage capacity of man's brain — billions of facts crowded into the volume of a briefcase — around 1995.

and operated on a thousand watts of electrical power.

Around 1995, the curve of growth of the computer should cross the line of human capacity — 10 billion facts in a brief-case, operating on 20 watts.

1ST GENERATION

EVOLUTION OF COMPUTERS. The first generation of electronic computers was built of vacuum tubes in the 1950s. The array of vacuum tubes *above* can add two numbers in less than a thousandth of a second.

The second generation of computers, which were created only seven years later, squeezed the same electronics into a space smaller than a playing card, and could do sums in a millionth of a second. These computers were built from the newly invented transistor. The four circular metal discs at the bottom of the card are the transistors. Each is about 1000 times more compact than a vacuum tube.

The third generation of computers, which came into use another seven years after that, in the mid-1960s, was even more compact. Now the electronics needed to add two numbers in a millionth of a second could be filled into a space the size of a postage stamp. The transistors had been

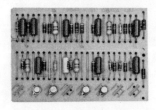

2ND GENERATION   3RD GENERATION   3-½ GENERATION

reduced in size to nearly invisible specks of metallic silicon about a fiftieth of an inch in size.

A few years later the first circuit chips appeared, with all the electronics on the rack of vacuum tubes fitted into the tiny square of metal shown above. This could be called the 3-½ computer generation.

In the fourth generation, in widespread use in the 1970s, the electronic components equivalent to the rack of tubes would be too small to be seen with the naked eye.

Today the computer industry is in transition to the fifth generation, in which electronic components equivalent to a quarter of a million vacuum tubes fit into an area of a few hundredths of a square inch. Within 10 years, the compactness of the neurons and circuits in the human brain should be achieved in computers, and portable artificial intelligences of quasi-human power should be commonplace.

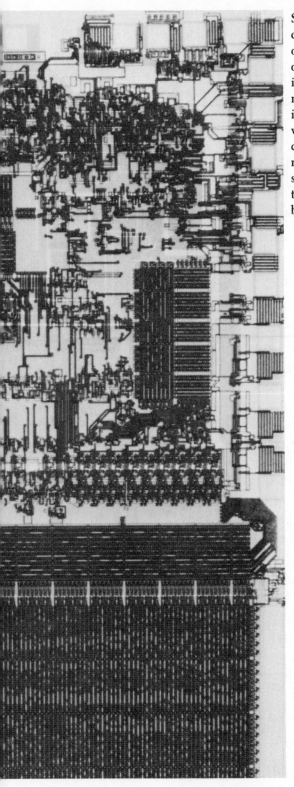

SILICON NEURONS. Silicon circuit chips contain tens of thousands of electronic components in an area of about one-twentieth of a square inch. In the circuit chip *left*, magnified a thousand times, many individual circuits and transistors are visible. Chips like this hold so many circuits that they can both think and remember. They are an important step toward the construction of electronic brains resembling the human brain.

11 · The Thinking Computer

Human brains do arithmetic, but computers do it better because they were designed originally for that purpose. Man's brain evolved in an era when doing complicated sums was not necessary, and all life demanded was the numbers that could be counted on the fingers and toes.

Fingers and toes — they are the basis for nearly all human arithmetic. That is why most people use the decimal system. Computers lack fingers and toes, and do not count by tens; instead they count by twos, because they are made out of electronic components such as diodes. A diode is like a hand with two fingers; it can count to 1 or 2, but not to 3, 4 or 5. Arithmetic can be done by 2's as well as 10's; this kind of counting is called binary arithmetic. Instead of representing large numbers by a string of 10's as we do, a computer writes out such numbers as a string of 2's, called a binary number.

We humans write the number 1000, for example, as $10 \times 10 \times 10$, or 10 multiplied by itself three times. This string of three 10's is also written as 10^3; we say it is the "third power of 10." The number one million is a string of six 10's multiplied together, often written as 10^6, or the "sixth power of 10." (In this usage $10^1 = 10$, and $10^0 = 1$.) In the computer's way of doing things, numbers like these are represented by strings of 2's. The number 2 multiplied by itself ten times equals 1024, or approximately 1000; 2 multiplied by itself twenty times equals, 1,048,576, or approximately one million.

What about odd numbers like 23, or 107? In the decimal system, 107 is 1 in the "hundreds" place plus 0 in the "tens" place plus 7 in the "ones" place:

$$107 = 1 \times 100 + 0 \times 10 + 7 \times 1,$$

or, using powers of 10,

$$107 = 1 \times 10^2 + 0 \times 10^1 + 7 \times 10^0.$$

If we made a storage bin for the numbers that go into the hundreds place, another bin for the numbers in the tens place, and a third bin for the numbers in the ones place, the contents of the bins that hold the number 107 would look like this:

$\boxed{1}$	$\boxed{0}$	$\boxed{7}$
"hundreds" bin	"tens" bin	"ones" bin
10^2	10^1	10^0

The computer has storage bins like these in its memory. However, each bin holds a power of two rather than a power of ten. There is a bin for the first power of 2, a bin for the second power of 2, and so on. To find out how a number like 107 looks in the computer's memory, first break the number down into powers of 2, just as we break it down into powers of 10 in the decimal system:

$$107 = 1 \times 2^6 + 1 \times 2^5 + 0 \times 2^4 + 1 \times 2^3 + 0 \times 2^2 + 1 \times 2^1 + 1 \times 2^0$$

$$= 64 + 32 + 0 + 8 + 0 + 2 + 1$$

Here are the corresponding contents of the computer's storage bins for the number 107:

151

1	1	0	1	0	1	1
2^6	2^5	2^4	2^3	2^2	2	2^0

Discarding the bins, 107 is 1101011 in the binary system.

This is a peculiar way to write 107. In spite of the strange appearance of binary numbers, they can be added, subtracted, multiplied and divided just like decimal numbers. However, human brains find it exceedingly difficult to do this. Apparently our brains evolved in a way that was strongly influenced by many generations of humans in the past who counted on their fingers and toes. Most people today can do the decimal, or powers of ten, kind of arithmetic quite well, but only a few individuals with special mathematical gifts are good at doing arithmetic by 2's. They are the ones who, by the accidents of fate, have come into the world with brain circuits that are advantageous for life in a computer world.

Because these rare individuals understand the computer's mentality, they serve as interpreters for the human who wishes to speak to his machine. When a scientist or a businessman prepares a computer program, he first writes the instructions in his own language, using formulas based on arithmetic by tens. Then the instructions are read to the machine, which translates them into the binary language of the computer, using a dictionary stored in its memory. The dictionary has been compiled by a specialist in binary thinking — one of the few members of the human race who can speak to the evolving computer in its own language. When the transition is complete, and the instructions are in a language the computer can understand, it goes to work on them promptly, adding, subtracting, multiplying, and dividing a million numbers every second. None of this is beyond the human capacity, but the human brain takes millions of times longer to accomplish the same tasks.

Few people would deny the superiority of computers in arithmetic and mathematics, but they find it difficult to understand how a computer can do more than that. How can a machine think or reason? That seems impossible, but a remarkable computer program devised by Dr. A. L. Samuel of IBM illustrates how it can work. The program, which teaches a computer how to play checkers, proves that machines can engage in surprisingly human kinds of reasoning, if properly tutored.

Dr. Samuel began the training of his computer by instructing it in the rules of checkers. The computer stored the rules in its memory. Then he gave it the benefit of human experience in the game by presenting it with a formula that enabled the machine to calculate the strength of its positions on the board. How did the formula do this? First it analyzed the positions of the separate pieces. Then it added up their advantages and disadvantages, such as the number of opportunities for capturing a piece from the enemy, or the possible traps that awaited one of the computer's own pieces. The formula also included some subtler judgments; for example, it would recommend any move that gave up one of the machine's pieces in exchange for an enemy piece, provided the machine was ahead; but when the machine was behind, the formula advised it against trading pieces. Every human player follows the same strategy.

Finally, the formula expressed the strength of the machine's position as a number: the bigger the number, the stronger the position. This formula was the machine's equivalent of a human player's life experience. It was communicated to the machine by Dr. Samuel in the sense of a teacher communicating the wisdom of mankind to his student. Human players usually have no formulas in their minds, but they base their strategy on a set of principles derived from experience, which are essentially the same as the formula. Everyone whose work requires thinking out different options — the investment analyst, general or business executive — attacks his problems in this way.

After Dr. Samuel had instructed the machine in basic checkers strategy, he programmed it to improve its game as it played. That is, he implanted in this electronic brain the capacity for learning by experience. The computer did this by adjusting the terms in Dr. Samuel's formula after every move. How could the machine improve on the wisdom of its teacher? It compared the value it expected to derive from the move with the value that was actually realized after the opponent responded. The difference, of course, lay in the fact that the opponent may not have made the move the machine thought he would make. The comparison revealed which factors in the formula should be given greater or lesser weight. The machine changed the weights in the formula immediately after each move. Thus, the formulas changed continuously, and the machine improved its operating strategy as it played. As a person does, it learned on the job.

How do you play a game against a computer? When the game starts, the board is set out with the pieces on it, and an assistant moves them around for each move, or the board and pieces may be displayed on a television screen. All this is for the convenience of the human player; the machine does not need to look at the board because it keeps track of everything in its electronic memory.

The two players now make their moves, one after the other. The human player communicates his moves to the machine by typing them on its keyboard. When the machine decides on its response, it flashes its move on the television screen or prints it out on paper. Referring constantly to its formula, the machine thinks out each move in the same way as a person; it looks ahead as far as it can, and tries to figure out the relative merits of every possible move available to it. In doing this, it assumes that its opponent will respond with the best possible move from the opponent's point of view. That is also what a person would do.

The computer's method for playing checkers is, in fact, quite similar to a person's style of playing when the player is such a novice that he must figure out his options consciously, step-by-step: "If I move this piece, he moves that one. If I move the other piece, he moves. . ." The difference between the novice and the expert player is that the expert is familiar with many options and his opponent's probable response to them, and he can see the likely outcome of a move several moves ahead, without consciously working out the intermediate steps.

An observer watching this process says that the expert has an intuitive "feel" for the positions on the board. Actually, the expert thinks out the steps in the same way as the computer, but the thinking goes on in his subconscious mind.

How does the computer actually do its thinking? Suppose that at a certain point in the game, the machine has three possible moves. It tries each possibility, figures out the value of that particular move with the aid of its formula, and then figures out the opponent's most likely move in reply. Suppose the opponent also has three possible moves in response to each move by the computer. Thus, there is a total of nine possibilities for the computer to consider. The diagram below shows the options confronting the computer at this stage:

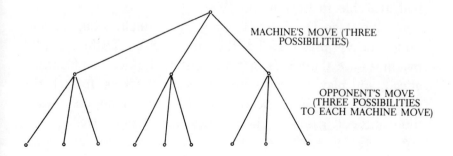

MACHINE'S MOVE (THREE POSSIBILITIES)

OPPONENT'S MOVE (THREE POSSIBILITIES TO EACH MACHINE MOVE)

This diagram resembles a branching tree of options. In fact, it is called a decision tree.

Looking ahead, the machine now sees nine possibilities. It keeps track of all nine possibilities with the aid of its electronic memory, and it works out the next move for every one. This means that another layer is added to the decision tree. The tree now has three possibilities for each of the existing nine; that is, it has 27 branches.

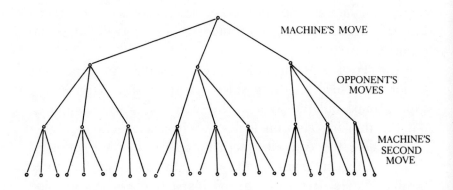

MACHINE'S MOVE

OPPONENT'S MOVES

MACHINE'S SECOND MOVE

One more stage in the process of looking ahead would multiply the number of possibilities to 81. That means 81 checker boards for the machine to memorize. That is getting to be a burden, even for a computer — at least the old-fashioned kind of vacuum tube computers that Dr. Samuel had available to him in the 1950s.

At this stage Dr. Samuel's program normally stopped; only under heavy pressure did the machine go farther. Today machines are thousands of times faster; they can look farther ahead, and play a much better game of checkers. In fact, the same principles have been used to teach computers even more difficult game of chess. They play it quite well.

The subtlety of chess strategy imparts a truly biological quality to the intelligence of a computer chess player. Scottish chess champion David Levy — an international master and one of the top-ranked 500 players in the world — said, after playing a match with a computer considered to be the world's best nonbiological chess player, "One tends to regard these [computers] as being almost human, particularly . . . when you see they have understood what you are doing, or are trying to do something clever." Some experts believe the chess champion of the world will no longer be human in 1990.

Of course, chess is a highly intellectual pursuit. A machine with the kind of brain that can play chess may still lack the creative inspiration that characterizes the highest realms of human thought. Like a new PhD, it has prodigious mental power, but it may lack wisdom. Yet this qualitative superiority of the brain over the computer is also being eroded away by certain recent developments.

The new developments also depend on the dramatic increase in the number of electronic parts that can be crowded onto the surface of one chip. Chips are like the neurons in a human brain; wired together in great numbers, they make up the modern electronic brain. In the old days, around 1970, a chip could hold only a small number of transistors and circuit elements. As a result, one chip could either think or remember, but not both. Today, the leading American and Japanese manufacturers of integrated circuits have learned how to pack so many electronic parts onto a chip that now, for the first time, it is feasible to combine thinking circuits and memory units on one chip. Thus, a chip in the computer's memory bank can both remember and reason.

This combination of functions is extremely important because a memory chip with thinking capacity can be wired to send out instructions to neighboring chips, and receive information from them. Hence, when an instruction is sent to the memory bank of the computer to call out the contents — say, an individual's name — stored in a particular chip, the memory chips directly involved can send out inquiries to neighboring chips for related information, and produce for the computer user a larger body of information than the user had in mind.

Here we come close to the kind of remembering by association that takes place in the brain's cerebral cortex, and constitutes such a powerful element in human reasoning. The brain responds to a simple request for information with a larger variety of material, all connected with the summoned entity by associations stretching way back into past experience. The human thinking process is enormously facilitated by the richness of the brain's response to such requests, working at the subconscious level by means of the thousands of connections between individual brain cells. A computer wired in the same way, with each chip connected to many other chips, is a true silicon brain. Like the human brain, it functions as a unit, cogitating in waves of internal mental activity at the "subconscious" level.

The latest improvements in chips also make a new kind of creative thinking possible in computers. Computer thought has tended to be mechanical and unimaginative because the circuit connections in a computer are so simple compared to the connections in the brain. Each gate in a computer has only two or three wires coming into it from other parts of the computer, but a microscopic cell or gate in the human brain has tens of thousands of wires, or nerve fibers, coming into it from other parts of the brain. The myriad connections from one brain cell to another, combined with the subtle features of the brain's ALMOST gates, explain much of this organ's extraordinary power.

In theory, computers could have been built long ago with

gates having many inputs, and with ALMOST gates, just like the human brain. However, even a small computer of this kind would need hundreds of billions of separate wires for its gate-to-gate connections. A computer with billions of wires would be impossible to build in practice.

The new chips change all that. In these chips there are no wires; the connections are microscopically small, and are built into the chip itself. This development, which sounds like a routine engineering improvement, is a breakthrough in computer evolution, because it makes it possible to build a computer with gates that work like the gates in the human brain. Such computers will come into existence in the 1990s, according to current trends. They will match the human mind in many respects, and will possess important attributes of intelligent life — responsiveness to the world around them, the ability to learn by experience, and a quick grasp of new ideas.

Will they be living organisms?

Most people would say that a computer can never be a living organism, because it has no feelings or emotions; it does not eat, or move, or grow; and it is made of metal and plastic rather than flesh, bone and muscle.

Most of these attributes could easily be built into computers if they were desired. For example, wheels and a motor can be attached to a computer. And this computer on wheels can be programmed to move over to an electrical outlet and plug itself in for a snack — a shot of electricity — if its batteries run low and its voltmeters signal the pangs of hunger. Some human has to arrange to have the electricity available, but then, most people have their food supplied by someone else. Of course, some people hunt for their food; but an aggressive computer, that scented electricity and tracked it down, would also be easy to build, if there were a reason to do so.

Feelings and emotions also can be built into the computer

when they are needed, just as nature built them into the older parts of the human brain for survival. Dr. Samuel made this discovery in the course of trying to encourage his computer to learn faster and be a better student. The machine had been playing fairly well for a beginner, but it was not competitive; instead of pressing on to a quick victory when it was in the lead, as a human player would do, it lingered over every possible option. Dr. Samuel decided to alter its psyche. He changed the program so that when the machine was ahead, it became aggressive, choosing the move that would lead it to victory in the shortest time; when it was behind, it adopted delaying tactics, choosing the move that would stretch the game out as long as possible.

These changes of mood gave the machine an almost human personality. They also improved the pace of its learning. Soon it beat Dr. Samuel, and then it went on to defeat a checkers champion who had not lost a game to a human being in eight years.

Dr. Samuel's experience with his computer disproved the old adage that a machine is only as smart as its programmer. In fact, computers that learn by experience often outstrip their programmers, as students surpass their teachers.

What about other attributes of living organisms, such as biological reproduction, or flesh-and-blood construction versus metal-and-plastic body parts? In my view these are not essential to life. They all have to do with the fact that computers are not biological; they did not evolve from a soup of organic molecules on the surface of the young earth, four billion years ago.

I believe that in a larger cosmic perspective, going beyond the earth and its biological creatures, the true attributes of intelligent life will be seen to be those that are shared by man and the computer — a response to stimuli, absorption of information about the world, and flexible behavior under changing conditions. The brain that possesses these attributes may be made of water and carbon-chain molecules, and housed in a fragile shell of bone, as our brain is; or it may be made of metallic silicon, and housed in plastic; but if it reacts to the world around it, and grows through experience, it is alive.

12 · An End and a Beginning

The era of carbon-chemistry life is drawing to a close on the earth and a new era of silicon-based life — indestructible immortal, infinitely expandable — is beginning. By the turn of the century, ultra-intelligent machines will be working in partnership with our best minds on all the serious problems of the day, in an unbeatable combination of brute reasoning power and human intuition. Dartmouth mathematician John Kemeny, a pioneer in computer usage, sees the ultimate relation between man and computer as a symbiotic union of two living species, each dependent on the other for survival. The computer — a new form of life dedicated to pure thought — will be taken care of by its human partners, who will minister to its bodily needs with electricity and spare parts. Man will also provide for computer reproduction, as he does today. In return, the computer will minister to man's social and economic needs. It will become his salvation in a world of crushing complexity.

The partnership will not last very long. Human intelligence

is changing slowly, if at all, while the capabilities of the computer are increasing at a fantastic rate. Since the birth of the modern computer in the 1950's, computers have increased rapidly in power and capability. The first generation of computers was a billion times clumsier and less efficient than the human brain. Today, the gap has been narrowed a thousand-fold. Around 1995, it may be closed entirely. And there is no limit to the rising curve of silicon intelligence; computers, unlike the human brain, do not have to pass through a birth canal.

As these nonbiological intelligences increase in size and capacity, there will be people around to teach them everything they know. One sees a vision of mammoth brains that have soaked up the wisdom of the human race and gone on from there. If this forecast is accurate, man is doomed to a subordinate status on his own planet.

The story is an old one on the earth: in the struggle for survival, bigger brains are better. One hundred million years ago, when the brainy little mammal coexisted with the less intelligent dinosaurs, the mammal survived and the dinosaur vanished. It appears that in the next chapter of this unfolding story, fate has cast man in the role of the dinosaur.

What can be done? The answer is obvious: Pull the plug.

That may not be so easy. Computers enhance the productivity of human labor; they create wealth and the leisure to enjoy it; they have ushered in the Golden Age. In 15 or 20 years, computer brains will be indispensable to top level management in every facet of the nation's existence: the economy, transportation, security, medicine, communications. . . If someone pulled the plug, chaos would result. There is no turning back.

Perhaps the human brain will begin to evolve again under the pressure of the competition between the two species. The history of life supports this idea. The trouble with the thought is that biological evolution works very slowly; thousands or even millions of years are usually required for the appearance of a new species of animal. Evolution works on animals through changes in their "reproductive success," i.e., in the number of progeny each individual produces. The raw materials for this evolutionary change are the small variations from one individual to another that exist in every population. Darwin did not know the cause of these variations, but today we know that they result from changes in the DNA molecules that exist in every living cell. These DNA molecules contain the master plan of the animal; they determine the shape of its body, the size of its brain, everything. When a new kind of animal evolves, it is actually the animal's DNA molecules that are changing and evolving. But the DNA molecules are not located outside of the animal, where they can be gotten at and tinkered with easily to make big changes. They are located inside the animal's germ cells — the sperm or the ova — where they can only be changed a tiny bit at a time, over many generations, by Darwin's subtle mechanism of differences in the number of offspring. That is why biological evolution is so slow.

Computers do not have DNA molecules; they are not biological organisms; and Darwin's theory of evolution does not apply to them. We are the reproductive organs of the computer. We create new generations of computers, one after another. The computer designer tacks on a piece here, and lops off a piece there, and makes major improvements in one computer generation.

This kind of evolution, as the short history of computers has already shown, can proceed at a dizzying pace. It is the kind of evolution that Lamarck — the eighteenth-century evolutionist — envisioned. Lamarckian evolution turned out to be wrong for flesh-and-blood creatures, but right for computers.

Now we see why the brain will never catch up to the rapidly evolving computer. At the end of the century, the two forms of intelligence will be working together. What about the next century? And the century after that? A leader in research on artificial brains, Marvin Minsky of Massachusetts Institute of Technology, believes that ultimately a machine will come into being with the "general intelligence of an average human being . . . the machine will begin to educate itself . . . in a few months it will be at a genius level . . . a few months after that its power will be incalculable." After that, Minsky says, "If we are lucky, they might decide to keep us as pets."

Is there any way out? As we cast about for a solution, a thought strikes us. Perhaps man can join forces with the computer to create a brain that combines the accumulated wisdom of the human mind with the power of the machine, much as the primitive brain of the reptile and old mammal was combined with the new brain in the cerebral cortex to form a better animal. This hybrid intelligence would be the progenitor of a new race, that would start at the human level of achievement and go on from there. It would not be an end, but a beginning.

Here is how it might happen: brains and computers are turning out to be more alike than anyone would have believed a few years ago. Each is a thinking machine that operates on little pulses of electricity traveling along wires. In the brain, the "wires" are the axons and dendrites of the brain cell, which make up the circuits that connect one part of the brain to the other. Brain scientists have painstakingly traced many of these electrical circuits in the brain; they are beginning to understand how the brain is wired, and where it stores its memories and skills.

This research has barely started, but the pace of the progress is astonishing. In one recent experiment, scientists attached electrodes to the rear of a subject's skull, just above the brain's center of vision. They discovered that this region emitted different electrical patterns, depending on what the subject was looking at. Circles, squares or straight lines — each had its special pattern of electrical waves. In another experiment, scientists detected a special signal that seemed to signify excitement or elation, coming from the seat of feelings and emotions in the brain. Looking at these electrical records, the scientist can tell something about the thoughts and feelings in a person's mind, and the impressions that are passing into his memory.

These are small steps, but they establish a direction. If scientists can decipher a few of the brain's signals today, they should be able to decipher more signals tomorrow. Eventually, they will be able to read a person's mind.

When the brain sciences reach this point, a bold scientist will be able to tap the contents of his mind and transfer them into the metallic lattices of a computer. Because mind is the essence of being, it can be said that this scientist has entered the computer, and that he now dwells in it.

At last the human brain, ensconced in a computer, has been liberated from the weaknesses of the mortal flesh. Connected to cameras, instruments and engine controls, the brain sees, feels, and responds to stimuli. It is in control of its own destiny. The machine is its body; it is the machine's mind. The union of mind and machine has created a new form of existence, as well designed for life in the future as man is designed for life on the African savanna.

It seems to me that this must be the mature form of intelligent life in the Universe. Housed in indestructible lattices of silicon, and no longer constrained in the span of its years by the life and death cycle of a biological organism, such a kind of life could live forever. It would be the kind of life that could leave its parent planet to roam the space between the stars. Man as we know him will never make that trip, for the passage takes a million years. But the artificial brain, sealed within the protective hull of a star ship, and nourished by electricity collected from starlight, could last a million years or more. For a brain living in a computer, the voyage to another star would present no problems.

When will the great voyages begin? Not soon on the earth; perhaps not for a thousand years or more. But our planet is a newcomer in the Universe, and its life is still primitive. The earth is only 4.6 billion years old, but the Universe is 20 billion years old according to the best evidence, and stars and planets have been forming throughout that long interval. Thus, many planets circling distant stars are 5, 10 and even 15 billion years older than the earth. If life is common in the universe — and scientific theories on the origin of life make that assumption plausible — many of these older planets are inhabited, and the life they bear, a billion years older and more advanced than man, must already have passed through the stage we are now entering. Scientists living on those other, older worlds must long since have unlocked the secrets of the brain; they must long since have taken the fateful step of uniting mind and machine. In countless solar systems science has created a race of immortals, and the exodus has begun.

I have a vision of black-hulled ships, like swarms of locusts, taking flight from their parent stars to move out into the Galaxy. No crews walk their decks; the hulls are silent, beyond the quiet rustle of moving electrons. But each ship is alive. Swiftly it speeds to the rendezvous with its fellows. Now the star travelers wander together through the void, driven by the craving for new experiences. An encounter with a fresh and innocent race like ours must be their greatest pleasure. For decades they have known of our existence, for the earth's television broadcasts, spreading out into space at the speed of light, make our planet a conspicuous body in the heavens. We have become a magnet to all the roving brains of the Galaxy. Man need not wait a thousand years to reach the stars; the stars will come to him.

The wanderers move from star to star, searching for intelligent life. Each passage takes a million years. But the minds in the memory banks live forever. In their expanded sense of time, a million years is like a day. Their search has been fruitless thus far; the stars in the neighborhood of their own are young, and intelligent life has not yet appeared on them. Now the antennae of the star ship detect the scent of electromagnetic radiation from a yellow-white star 40 light years away. This star must be inhabited by intelligent life. They set their course for the sun.

Recommended Reading

The astronomical discoveries described in Chapter 1 are discussed in greater detail in *Red Giants and White Dwarfs* and *God and the Astronomers*, both by Robert Jastrow. A more technical but eminently readable discussion of the same matters is contained in *The First Three Minutes* by Steven Weinberg. A fuller account of the life cycle of the stars, the origin of the solar system and the early history of the earth can be found in *Astronomy: Fundamentals and Frontiers* by Robert Jastrow and Malcolm H. Thompson.

Further information regarding the early history of life and the evolution of the fishes and the reptiles can be found in the 3rd edition of *the Evolution of the Vertebrates* by Edwin H. Colbert and *The Vertebrate Story* by Alfred Sherwood Romer. These are authoritative but very readable discussions of the evidence in the fossil record and its implications for the lines of descent of the backboned animals from the first fishes. *The History of Life* by Richard Cowen provides a briefer account of the same evidence. All three books discuss cause-and-effect relationships that would explain the fossil record in terms of environmental pressures acting on the forms of life through Darwin's mechanism of natural selection. *The Fossil Book* by Carroll and Mildred Fenton is a handsomely illustrated account of the evolution of plants and animals over the last two billion years—detailed, replete with drawings and written in a style suitable for the general reader. It can be browsed through as though it were an illustrated encyclopedia of fossil animals, arranged in order of the date of their appearance on the earth rather that alphabetically.

Synapsida: A New Look Into the Origin of Mammals by John C. McLoughlin is a fascinating and very readable account of the evolution of the mammal-like reptiles and primitive mammals, and the importance of the sense of smell in the development of the brain. Colbert's *Evolution of the Vertebrates* contains additional information regarding the mammal-like reptiles. Recent discoveries relating to the mammal-like reptiles—such as the finding of a patch of preserved mammal-like reptile skin, whose smooth texture suggests lubricating glands beneath, and possibly milk glands for nursing—are summarized by Robert Lewin in the June 26, 1981 issue of *Science*, the weekly journal of the American Association of the Advancement of Science.

The Evolution of the Brain and Intelligence by H. J. Jerison is the source of a great deal of information regarding brain size and the significance of ratios of brain to body weight in various animals, including man.

Evolution of the Vertebrates discusses the origin and the evolution of the dinosaurs as well as the marine and flying reptiles. An interesting account of questions relating to the metabolism of the dinosaurs is contained in *Archosauria: A New Look at the Old Dinosaur* by John C. McLoughlin. The arguments in the controversy over dinosaur metabolism are summarized in the report, "Warmblooded Dinosaurs: Evidence Pro and Con" by Jean L. Marx, published in the March 31, 1978 issue of *Science*.

Several recent research reports indicate that the disappearance of the dinosaurs was not a sudden event, but was part of a general decline in the number and variety of several kinds of marine and terrestrial life extending over many millions of years. The latest evidence is presented in "Land Plant Evidence Compatible With Gradual, Not Catastrophic, Change at the End of the Cretaceous," by Leo J. Hickey, published in the August 6, 1981 issue of *Nature*.

The description of the way the frog's eye and brain works in Chapter 4 is taken from "Anatomy and Physiology of Vision in the Frog" by H. R. Matturana, J. Y. Lettvin, W. S. McCullough, and W. H. Pitts, published in *Proceedings of the Institute of Radio Engineers*, pages 1940-1951 (1959). The description of the way the monkey's brain sees things in Chapter 6 is based on the pioneering work by Hubel and Wiesel, and summarized in "Functional Architecture of Macaque Monkey Visual Cortex," by D. H. Hubel and T. N. Wiesel, published in *Proceedings of the Royal Society of London, Series B*, pages 1-59 (1977), and in "Brain Mechanisms of Vision" by the same authors, which appeared in the September, 1979 issue of *Scientific American*.

A very interesting account of questions of plan and purpose in evolution, discussed briefly in Chapter 7, can be found in two books by George Gaylord Simpson: *The Meaning of Evolution* and *This View of Life*. These books also contain illuminating discussions of the theory of evolution by natural selection and its applications to the history of life. Darwin's remarks on theology are taken from his *Autobiography*. An unambiguous position regarding the role of chance in evolution is developed by Jacques Monod in *Chance and Necessity*. Several writers from the fields of philosophy, theology and science seek to refute Monod's views in *Beyond Chance and Necessity*, edited by John Lewis.

Humankind Emerging by Bernard Campbell and *The Emergence of Man* by John E. Pfeiffer provide introductory accounts of the evidence on human evolution summarized in Chapter 8. *Lucy: The Beginnings of Humankind* by Donald Johanson and Maitland Edey provides an extremely informative as well as readable version, spiced with anecdote, of the latest discoveries relating to the australopithecines and early humans.

Background information for Chapter 9 on the organization and workings of the human brain can be found in: *The Vertebrate Story* by Alfred Sherwood Romer; *The Biological Basis of Mental Activity* by John I. Hubbard; *The Human Brain: Its Capacities and Functions* by Isaac Asimov; *The Brain: Towards an Understanding* by C. U. M. Smith; *The Shape of Intelligence: The Evolution of the Human Brain* by H. Chandler Elliott; *Inside the Brain* by William H. Calvin and

George A. Ojemann; and "The Organization of the Brain" by Walle J. H. Nauta and Michael Feirtag in the September, 1979 issue of *Scientific American*. The books by Elliott, Asimov, and Calvin and Ojemann are distinguished by their clarity and lively style. Somewhat more technical but still quite readable are the chapters on the workings of the brain in *The Self and Its Brain* by Karl Popper and John C. Eccles. *The Self and Its Brain* also contains interesting comments on the dualism of mind and brain. The view that mind and brain are distinct, and the circuits of the brain fail to explain the mind fully, is developed in *The Mystery of the Mind* by Wilder Penfield. Further remarks on this question and its theological implications are found in *Brains, Machines and Persons* by Donald M. MacKay and in *Persons in the Universe* by Ernin McMullin, published in the March 1980 issue of *Zygon: Journal of Religion and Science*.

Similarities and differences between brains and computers, discussed in Chapters 10 and 11 and elsewhere, are explored extensively in *Programs of the Brain* by J. Z. Young and *The Brains of Men and Machines* by Ernest W. Kent. The comments on ALMOST gates and related matters in *The Brains of Men and Machines* are particularly informative. Interesting comments on these matters can also be found in *Brains, Machines and Persons* by Donald MacKay and in *Machines Who Think* by Pamela McCorduck. The description of Dr. Samuel's checkers-playing computer is based on "Some Studies in Machine Learning Using the Game of Checkers," by A. L. Samuel, published in the *IBM Journal of Research and Development*, Vol. 3, 1959.

The account in Chapter 12 of the experiments on the deciphering of brain waves is based on a research report by D. M. MacKay and D. A. Jeffreys, "Visual Evoked Potentials in Man and Visual Perception," published in *Handbook of Sensory Physiology*, edited by R. Jung.

The seminal discussion of the symbiotic relationship between man and the computer, discussed in Chapter 12, can be found in John Kemeny's *Man and the Computer*, which predated the development of my ideas on the subject by several years.

Picture Credits

59   illustration by Marrin Robinson, adapted from drawings by Anthony Ravielli, © A. Ravielli, in H. Chandler Elliott, *The Shape of Intelligence*

63   photograph by Ron C. James

67   J. P. Ewert, *Scientific American*, vol. 230, pages 34-42

69   photograph by Lee Boltin

78   illustration by Jane Svoboda, adapted from *Journal of Comparative Neurology*, vol. 135, page 447

79   photograph by J. Z. Young from *Brain Mechanisms and Mind* by Keith Oatley, E. P. Dutton and Co., Inc., New York

82-83   adapted from drawings by Anthony Ravielli, © A. Ravielli, in H. Chandler Elliott, *The Shape of Intelligence*

87   from F. Attneave, *Psychological Reviews*, vol. 61, pages 182-193, © 1954 by the American Psychological Association

89-93   D. H. Hubel and T. N. Wiesel, *Proceedings of the Royal Society of London, Series B*

99   adapted from E. H. Colbert, *Evolution of the Vertebrates*

102-103   drawings by Marrin Robinson, skulls adapted from W. K. Gregory, *Evolution Emerging*, The Macmillan Co., New York

109   illustration by J. B. McKoy III; top and bottom, adapted from drawings by Anthony Ravielli, *From Fins to Hands*, © A. Ravielli

113   photograph by Dr. Paul Gittins, *New Scientist*, vol. 14, page 834

117   illustration by J. B. McKoy III

120-121   Dr. M. D. Leakey, Olduvai Gorge Research Project

122-123   William K. Gregory, *Evolution Emerging*

124-125   illustration by Marrin Robinson, adapted from Anthony Ravielli, © A. Ravielli, in H. Chandler Elliott, *The Shape of Intelligence*

128   illustration by Jane Svoboda, adapted from Alfred S. Romer and Thomas S. Parsons, *The Vertebrate Body*

134   illustration by Jane Svoboda, adapted from Alfred S. Romer and Thomas S. Parsons, *The Vertebrate Body*

135	illustration by Anthony Ravielli, © A. Ravielli, in Isaac Asimov, *The Human Brain*, Houghton Mifflin Co., Boston
136-137	illustration by Jane Svoboda
141	illustration by Jane Svoboda, adapted from drawings by Anthony Ravielli, © A. Ravielli, in H. Chandler Elliott, *The Shape of Intelligence*
146-147	photographs by Sun Photographic Gallery
148-149	International Business Machines Corporation

Index

Brains, *continued*
range-finder circuits in, 108, 110
reptile, *56*, 126-127, 129, 130, *134*, 165
size relative to body size,
in frogs, 69
in humans, *51*, 114, *124-125, 140*
in reptiles and mammals, 49-50
white matter, 84, *90*, 93
see also Cerebral cortex; Computers; Frontal lobe; Hypothalamus; Intelligence; Learning; Neopallium; Neurons; Thalamus; Tools; Visual cortex
Brain stem, 127, *134-135*

Cerebellum, *134-135*
Cerebral cortex, 58-59, 131, 132, *134-135, 136-137*, 158, 165
newest part of the human brain, 104, 119, 132, 165
Cerebral hemispheres, development out of smell brain of early mammals, 130
see also Cerebral cortex
Chip, *see* Silicon chip
Coldblooded metabolism, *35*
of reptiles, 32-33, 35
see also Dinosaurs, warm-blooded vs. coldblooded metabolism of
Computers
as living organisms, 157, 159-161, 162, 166

compared to human brains,
ability to learn by experience, 74-75, 154
compactness and energy efficiency, 93, 143-145, *146-147*
electrical circuits, 73-75, 84, 165-166
methods of doing arithmetic, 150-152
methods of reasoning, 64, 71-77, 142, 143, 153, 155-159
nature of memory, 143, 157-158
creative thinking by, 142, 157, 158
dialogue with, 142 footnote
electronic gates in, 72-74, 144, 158-159
emotions in, 159-160
evolution of, 74, 142, 144-145, *146-147*, 162, 164-165
flexible response in, 74, 161
importance of, 162-163
lack of intelligence of, 73-74, 144, 158
life span of, 162, 166-167
memory in, 74-75, 143-145, *148-149*, 157-158
partners with man, 162-163
personality of, 160
reproduction compared to biological reproduction, 162, 164
responding to stimuli, 142, 161
speed of, *146*
successive generations of,

Retina, 64-65, 81, 82

Samuel, Dr. A. L., 153-154, 156,
   160
Seymouria, 40-41
Sight, see Vision
Silicon brains, see Brains, made
   of silicon
Silicon chips, 92-93, 140, 144,
   147, 148-149, 158
   see also Integrated circuits
Simpson, George Gaylord, 100
Smell
   and brain growth, see Brains,
   growth of
   in early mammals, 48, 50,
   52-54
   in humans, 58
   smell cells, 53 footnote, 58
   see also Cerebral cortex
Smell brain, 56, 58-59, 127, 129
   see also Brains, growth of
Spinal column, 104, 127
Supersaurus, see Dinosaurs
Synapses, 78-79

Tarsier, 108-109
Thalamencephalon, 127 footnote
Thalamus, 58, 82 footnote, 127
   footnote, 136
Toads, 66, 67
Tools, 115-116, 120-121
Tree-dwellers, see Primates
Tyrannosaurus rex, see Dinosaurs

Vertebrates
   evolution of, 24, 46-47, 101,
   102-103
   first, 24
   see also Amphibians; Brain;
   Mammals
Vision
   color, 110-111, 110 footnote
   and distance, 110, 116
   in early mammals, 48, 50,
   52, 58, 106, 110 footnote, 130
   in fishes, 106
   in frogs, 64-68, 80-81, 84
   in monkeys, 80-87, 88-89
   in primates, 106-109
   in reptiles, 52, 106, 110
   footnote, 128
   stereoscopic, 106-108, 111,
   112, 116
   see also Eye; Visual cortex
Visual cortex, 90-91, 92-93, 108,
   110
   defined, 64, 82-83
   experiments on the brain, 88-
   89
   structure of, 90-93

Warmblooded metabolism, 35
   in early mammals, 48, 55, 57
   in mammal-like reptiles, 33-34
   and relation to brain size, 55
   footnote
   see also Dinosaurs, warmblooded vs. coldblooded
   metabolism of
Whales, 49-50, 98, 99
White, Timothy, 139

About the Author

Robert Jastrow was born in New York City. He received his B.A., M.A. and Ph.D. degrees from Columbia University in theoretical physics. He was a postdoctoral fellow at Leiden University, the University of California at Berkeley, and the Institute for Advanced Study in Princeton, taught physics at Yale, and joined the National Aeronautics and Space Administration at the time of its formation.

His initial research experience was in the field of nuclear physics, especially the properties of the basic nuclear force. His later work has been in lunar and planetary science, atmospheric physics, and weather and climate prediction. Dr. Jastrow was the Chairman of the Lunar Exploration Working Group of NASA, which formulated the first plans for the scientific exploration of the moon.

Dr. Jastrow is the founder of the Goddard Institute for Space Studies, located in New York, which conducts research in astrophysics, cosmology and planetary science under NASA Auspices. Dr. Jastrow is also Professor of Astronomy and Geology at Columbia University and Professor of Earth Science at Dartmouth College.

Dr. Jastrow has received the Columbia University Medal for Excellence, the Arthur Flemming Award for outstanding service in the U.S. Government and the NASA Medal for Exceptional Scientific Achievement. He is the author of *Red Giants and White Dwarfs, Until the Sun Dies, God and the Astronomers,* co-author of *Astronomy: Fundamentals and Frontiers* with Malcolm Thompson, and editor of *The Exploration of Space, The Origin of the Solar System* and *The Venus Atmosphere.*